CHANGING ORDER

H.M.COLLINS

Replication and Induction
n Scientific Practice

Changing Order

CHANGING ORDER
Replication and Induction in Scientific Practice

H.M. Collins

SAGE Publications · London · Beverly Hills · New Delhi

SAGE Publications Ltd
28 Banner Street
London EC1Y 8QE

 SAGE Publications Inc
275 South Beverly Drive
Beverly Hills, California 90212

SAGE Publications India Pvt Ltd
C-236 Defence Colony
New Delhi 110 024

British Library Cataloguing in Publication Data
Collins, H.M.
   Changing order: replication and induction in scientific practice.
   1. Science — Social aspects
   I. Title
   306'.45   Q175.5

Library of Congress Catalog Card Number   **85-040195**

ISBN 0-8039-9757-4
ISBN 0-8039-9717-5 pbk

Printed in Great Britain by J. W. Arrowsmith Ltd., Bristol

Contents

Preface and Acknowledgements

This book shows how ships get into bottles and how they get out again. The ships are bits of knowledge and the bottles are truth. Knowledge is like a ship because once it is in the bottle of truth it looks as though it must always have been there and it looks as though it could never get out again. Since order and knowledge are but two sides of the same coin, changing knowledge is changing order. This book takes scientific knowledge as a case study.

I began to work on these themes in the early 1970s but the book includes ideas and approaches that go back to the beginning of my education in sociology; my debts go back correspondingly far.[1] The greatest of these is to my friend and first teacher of sociology — Reg Hughes — who brought the subject alive for me. Later, I was lucky to encounter at the University of Essex a group of fellow students and staff who further encouraged curiosity about fundamental problems of philosophy and methodology. I then found among my colleagues at the University of Bath an encouragingly pluralist atmosphere, not to mention a saintly patience in the face of overweening enthusiasm.

As far as this particular book is concerned I would like to thank my friends at the Programme in the History of Science at Princeton University who provided both the mental and physical space for me to complete the first draft. I thank also Barry Barnes, Graham Cox, David Edge, Trevor Pinch, Steve Shapin, David Travis, David Gooding and Alice Leonard for reading and commenting on parts of the manuscript. The last two in particular worked well beyond the demands of regularly reciprocated duty and saved me from some serious philosophical and stylistic errors. The faults that remain are entirely my own.[2]

I also thank Bob Harrison for putting up with me as an amateur laser-builder for a number of years and for being a sociological victim. Bob Draper and Dick Metcalfe helped me with the experiments on the emotional life of plants described in Chapter Five.

Elizabeth Sherrard and Sandra Swaby typed a great deal of the first draft of the manuscript, struggling with a new word-processing package. Sage, the publishers, have acted with decency and despatch — by no means universal qualities in academic publishing — and Farrell Burnett's conscientious editing has been of great value. Finally, I thank Pat Ryan for words of comfort and for putting fewer obstacles in the way of the work than her principles demanded and our children, Joe and Lily, for remaining good-humoured throughout the ordeal.

University of Bath
September 1984

Notes

1. Readers who are familiar with my work will recognize many of the themes and ideas in the book from previously published papers. Nevertheless, there are very few pages that have been reproduced identically in their previously published form; such as there are will be found in the first parts of Chapter Three and some sections of Chapter Four. Otherwise, even familiar material has been broken up and re-written to fit with the story of the book. I thank the editors of *Social Studies of Science* for permisssion to reprint some pages from my articles 'The TEA-Set' (Collins, 1974), 'Building a TEA-Laser' (Collins and Harrison, 1975) and 'Son of Seven Sexes' (Collins 1981c) and the editor of *Sociology* for permission to reprint some pages from 'The Seven Sexes' (Collins 1975). Other material has appeared as unpublished conference papers — notably a paper on replication in parapsychology which is now distributed within chapters Two and Five and another on the function of calibration which is spread throughout Chapters Four and Five. Chapters One and Six, the Postscript, the approach of Chapter Two and the bulk of Chapter Three are almost entirely new and so is the framework of the entire argument. The idea of the 'experimenters' regress', around which the book turns, underlies nearly all of my earlier work, but had not previously been explained in sufficient detail.

2. The fieldwork reported in this book was supported by a grant from the Social Science Research Council (now the ESRC) and a small grant from the Nuffield Foundation.

Introduction

[In Tlon there are] objects composed of two terms, one of visual and another of auditory character: the colour of the rising sun and the far-away cry of a bird. There are objects of many terms: the sun and the water on a swimmer's chest, the vague tremulous rose colour we see with our eyes closed, the sensation of being carried along by a river and also by sleep ... (Jorge Luis Borges, 'Tlon, Uqbar, Orbis Tertius')

During the last decade sociologists, historians and philosophers have begun to examine science as a cultural activity rather than as the locus of certain knowledge. The ideas that come out of this research have significance for more than a few specialist academics, for when it is regarded in this way the study of science can tell us things about culture as a whole — while at the same time this new perspective demystifies the role of scientific expertise. The importance of these ideas both for professional academics and for all those whose lives are touched by science has led me to try to provide a readable account while still making a technical contribution. Thus I have attempted to write the main text in a way which makes it accessible to anyone with an interest in and some knowledge of either the natural sciences, social sciences, history of science or philosophy. I have also added a 'Postscript' which explicitly draws out the book's wider relevance.

To keep the text within bounds of length and comprehensibility I have made much use of discursive notes in some chapters, most of which would have been included in a longer text. I hope the interested but non-specialist reader will turn to at least some of them. A few of the notes are intended purely for the specialist and I have indicated this by putting the superscript in parentheses — thus.[56] I have also gathered some technical material that is not likely to be of interest to the general reader into a short methodological appendix.

The effort that the book does require is an initial derailment of the mind from the tracks of common sense. Our cultural environment — the everyday world — has to be turned into a strange place if we are to see that its perceived orderliness is a remarkable and mysterious human accomplishment. I have tried to engender this shift of consciousness in the reader by introducing the book with problems arising out of philosophical scepticism. The outcome is a form of 'relativism'— a term and a philosophy that frightens many. But this form of relativism is a pleasant glade not far from the perceptual railway tracks we normally ride on. Indeed, the relativist glade has paths in it that lead to most of the destinations of the metal road. However, they do not lead in quite the same predetermined way; a woodland track invites exploration, has alternate routes and offers a wider choice of scenery than a railway.

The first chapter has been designed to open the mind for an exploration of the fundamental problems of order in conceptual and social life. It shows that our concepts and our social conventions reinforce each other — as in a network — and this explains the maintenance of order. Concepts and conventions are 'jointly entrenched' within 'forms of life'. The problem of change is left for empirical investigation in later chapters. Chapter One concludes by showing how the fundamental problems of conceptual order and change give rise to the well known difficulties encountered in building intelligent machines.

The second chapter examines science's ordering principle — the replicability of observations and experiments. I adopt a metaphor from *The Hitch-Hikers Guide to the Galaxy*: the Earth as a computer constructed by 'philosopher-mice'. We are able to see the problems of intelligent machines, discussed in Chapter One, re-emerge when science as a whole is thought of as a giant computer. In particular, it does not seem possible to construct a computer-type 'algorithm' for ensuring that experimental replication always provides a definitive test for the existence of new and disputed natural phenomena.

Chapters Three, Four and Five report the main field studies. They are all close examinations of passages of science in which scientists tried to repeat each other's work: laser building — a relatively straightforward piece of science; the detection of gravitational radiation — an area at the very frontiers of research; and 'the secret life of plants' and 'mind-over-matter' — areas of parapsychology. Chapter Four concludes with a technical appendix on the detection of gravity waves.

The main argument of these three chapters turns on a comparison of the process of replication of scientific findings among these areas of science. This comparison will reveal the existence of what I have called the *experimenters' regress*. This is a paradox which arises for those who want to use replication as a test of the truth of scientific knowledge claims. The problem is that, since experimentation is a matter of skilful practice, it can never be clear whether a second experiment has been done sufficiently well to count as a check on the results of a first. Some further test is needed to test the quality of the experiment — and so forth.

Both Chapters Four and Five conclude with a discussion of ways in which scientists try to test the quality of experiments directly in order to circumvent the experimenters' regress. In Chapter Four the process of calibrating the apparatus is discussed; in Chapter Five it is the use of surrogate phenomena. The failure of these 'tests of tests' to resolve the difficulty demonstrates the need for further 'tests of tests of tests' and so on — a true regress.

The sixth chapter pulls together the themes of Chapters Two, Three, Four and Five and develops their implications for the problems of perceptual order set out in Chapter One. I show how the individual scientist is tied into the network of institutions in the wider society and try to demonstrate how these constrain research choices and the outcome of work at the laboratory bench. The sources of stability in the conceptual universe are explored and the problems and means of 'changing order' are discussed. During the discussion a number of propositions about experiment are developed. The first ten of these are available for easy reference on the first page of Chapter Six.

A Postscript discusses the wider implications of the book for science education, science policy questions, forensic science, public inquiries and the role of scientific expertise in the institutions of democratic society. It concludes with some examples of the way that an understanding of scientific change sheds light on political processes. A methodological appendix follows.

It might be thought that the passages of science compared in Chapters Three, Four and Five are selected according to a strange principle. After all, there is one piece of near-technology done without any intent to 'test' a finding — the TEA-laser; there is one piece of science drawn from a central theoretical tradition of physics, albeit one using frontier technology and coming up with some unexpected findings — the detection of gravity waves; and there are two pieces of parapsychology, one of which was done by what one might call the 'marginal man's marginal man', on the psychic life of plants. From a viewpoint which makes clear distinctions between 'real' science and 'pseudo' science, these would be odd things to *compare*. From the viewpoint of this book they are not such strange bedfellows; the relativist attitude demands that the analysis of the way that knowledge is established is not shackled at the outset by common sense judgements about what is and what is not true. The question is, rather, how things come to be seen as true or false; and this requires the self-conscious innocence which goes along with the suspension of everyday certainties.

These three examples of scientific practice are chosen for comparison because they represent two out of what I shall call the 'three phases' of science. These comprise the 'revolutionary' phase, the 'extraordinary' phase and the 'normal' phase. In the revolutionary phase large scale and widespread changes in the whole conceptual structure of disciplines take place. This idea, due to Kuhn (1962), has been the subject of heated philosophical debate; it is not discussed in this book (but see Collins and Pinch, 1982). The extraordinary phase, on the other hand, is easy to recognize. It is the site of smaller scale controversy. Such controversy arises when claims are made that do not sit comfortably with the prevailing orthodoxy. When the stormy waters settle, what is left is the normal phase (another, much less contentious, Kuhnian term) in which nearly all science is actually done. The case studies reported here are representative of the normal phase (TEA-lasers) and the extraordinary phase (gravity waves and parapsychology). The parapsychological study has, perhaps, proto-revolutionary qualities, and it is certainly a little further away from the centre of orthodoxy than even the gravity wave story.

As will be seen, the parapsychology and the gravity wave debates reported in this book look very like each other in terms of the structure of argument which surrounds the claims of replicability; and they both look quite unlike the TEA-laser case. Thus, if there is anything odd about comparing such a heterogeneous collection of passages of scientific activity, it is the TEA-laser that is the odd one out! This marks a difference between the perspective of this work and more orthodox ways of looking at science. In earlier perspectives it would be parapsychology that would look odd because of its marginality with respect to mainstream science, and because the other two cases are drawn from physics whereas it is a study of living subjects. But, as Chapters Three, Four and Five will reveal, the important dimension turns out not to be the scientific subject matter but the phase of science that is represented.

Chapter One

The Mystery of Perception and Order

A few years ago a sketch on television's *Monty Python's Flying Circus* featured a misleading Hungarian phrasebook. 'Can I have a box of matches?' was mistranslated into English along the lines of 'I would like to feel your beautiful thighs.' The appropriate rebuff in English was mistranslated into Hungarian as 'Your eyes are like liquid pools.' The phrasebook turned what should have been a routine exchange between a large Hungarian and a meek tobacconist into a violent brawl. The phrasebook introduced disorder into what the participants expected to be routine orderly interaction.

Without order there can be no society. Communication, and therefore the whole of culture in its broadest sense, rests on the ability of human beings to see the same things and respond to them in the same ways. There may be variation in perception and meaning between different groups, but the very existence of 'groups' depends on the uniformities within them. The fact is that there are groups, societies and cultures; therefore, there must be large scale uniformities of perception and meaning.

Though these uniformities are fundamental, the way that they come about and the way that they are maintained are profound mysteries. These mysteries underlie the major problems of philosophy, linguistics, sociology, artificial intelligence, and the philosophy of science. Concerted perception and understanding in an open environment seem to be something that humans simply do, without consciously thinking about it. This book is about the way such concerted perception and action come about. It explores the problem by focusing on the particular way scientists come to perceive, describe and understand new natural phenomena in a uniform way. It examines some instances of this sort of agreement and offers these instances as exemplary cases of the formation and maintenance of more general patterns of action. The field studies reported, narrowly focused as they are, are intended to reflect light on to the deeper problem of culture.

The difficulty is that, because we manage concerted perception and action with such unthinking ease, it hardly seems like an accomplishment of any note. Our common perceptions, as I have suggested elsewhere (Collins, 1975), are like ships in bottles. The ships, our pieces of knowledge about the world, seem so firmly lodged in their bottles of validity that it is hard to conceive that they could ever get out, or that an artful trick was required to get them in. Our world is full of ships already within their bottles and it is only the rare individual who gets a brief glimpse of the ship-in-bottle maker's art. Science, more than any other cultural activity, is in the business of putting new ships into new bottles — that is, it is in the business of building new bits of knowledge. Even in science, however, the art is so routinized that the tricks are only visible when some self-conscious attention is given to them, for instance, in the case of a scientific controversy. The first task, then, is to pull the mind free of the taken-for-granted ways of seeing and, instead, to let it see the sticks and strings and glue from which the ships of knowledge are built. I use philosophical scepticism, which is safe, legal and inexpensive, to loosen the trammels of commonsense perception.

Scepticism and the problem of inductive inference
Scepticism begins with the problem of why we should expect the future to be like the past. Why do we expect regular sequences of events to continue and how is it that we seem to obtain evidence about the future by extrapolating from past regularities? Inferring general rules from repeated past regular instances is called induction. Thus scepticism engenders what is known as 'the problem of induction'.[1] This is a philosophical problem having to do with the way that our 'inductive inferences' — generalizations from past experience — can ever be certain or even probable. At root, however, scepticism concerns the perception of any sort of regularity at all. My concern is not with how we could be certain *in principle* about induced regularities but about how we actually come to be certain about regularities *in practice*. This is a shift of focus which allows what I will call a 'sociological resolution' of the problem of induction. Many fundamental questions to do with the discovery of the rules of scientific and everyday activity are really locally specified versions of this practical puzzle.

The easiest way to begin to see this problem, and the standard starting point, is with the work of the philosopher David Hume. Hume posed the problem in terms of our ideas about cause: take an event — call it 'a' — such as the striking of a stationary billiard ball by another, and a second event — 'b' — the billiard ball moves across the table. We are inclined to say that the the billiard ball 'is propelled' across the table since we take it that its movement is 'caused' by the impact. We see sequences like a – b happen frequently and indeed we see and know *from experience* that the first billiard ball is what causes the second to move. But, suppose the regularity of the a – b sequence were just an extended coincidence, not a causal relationship — how would we see the difference? In other words, what is it that we see in the impact of the billiard balls that makes us view it as a causal relationship, which we are confident will continue, rather than an extended coincidence which we would not expect to continue? The answer is 'nothing'. Why then do we treat the relationship as one of necessity? Why do we think such relationships contain causal certainties?

One can sharpen the impact of the question by thinking of repeated sequences of events that we do not treat in this way. For example, we do not think that accurate forecasting of, say, the weather carries an implication of cause, even though in this case an event 'a', such as the vocalization of the words 'it will rain' is regularly followed by event 'b': precipitation. Of course, the weather forecast is not always right, but if it were to become more and more accurate we would not become more and more tempted to impute a causal relationship between the weather and the forecast. In the last resort, were the forecast to become wholly accurate, we would sooner be inclined to impute clairvoyance to the forecaster than causal efficacy to the forecast. Again, there are regular sequences which seem to incline us to think that the longer they carry on the more likely they are to break down. For example, naive roulette players will be more and more inclined to bet on black the longer an unbroken sequence of reds continues. Here they see Event A — the spin of the wheel — followed regularly by Event B — the falling of the ball into the red slot — and they become more and more certain that next time B will not follow A. These two examples show us that regularity of events in itself does not compel us to see causal relationships.

Hume thought that, although causes were invisible, we had a psychological propensity to impute necessity, and therefore cause, to regularly repeated sequences. No doubt there is truth in this; man is fundamentally a regularity-perceiving creature. We do, as a matter of fact, induce from the particular to the general all the time.[2] As Hume saw, however, this sort of 'solution' to the problem does not help because it would only be useful were it obvious beforehand which sequences of events could be generalized; as we have remarked, there are many sorts of regular sequences, such as the forecast and the weather, that we might generalize but don't, and many irregular ones that we do.[3] For these reasons, citing a generalizing propensity as an explanation of our inductive (regularizing) tendencies is simply truistic. Since a generalizing tendency would allow us to see anything and everything as regular, which would amount to seeing nothing, this sort of 'solution' is vacuous. In fact, the problem of perceiving regularity is a sub-division of the broader question about the very possibility of perception.

(The impatient reader may be wondering why scientific knowledge has not been brought in to rescue us from this artificial-looking problem. Surely it is science and its common sense counterparts which give us our warrant for thinking some sequences are properly thought of as regular and others as merely coincidental. Exactly so — this book is about how certain sequences and objects obtain their scientific *imprimatur*.)

Perception and stability of perception are the same thing. Try to imagine what disordered, chaotic perception would be like. We are looking at a red bus, the next moment it appears to be a tiger and growls; the next instant it looks and tastes like a lemon, the next like a black swan, and so on. Imagine such a sequence taking place randomly at great speed, and take this to be our normal way of seeing things. Under these circumstances we would not want to take a ride on a bus. Since we would never want to ride on a bus and, by extension, we would never have ridden on one, and neither would anyone else, no one would even know that buses were things one could ride on. This means that we would not have the concept of bus available to us, and this makes it nonsensical to talk of having seen a bus in the first place. The best we could have experienced would have been a kind of nondescript (literally) reddish thing. Thus, even to perceive the reddish thing as a *bus* requires that we, and the driver and conductor, expect the bus to stay a bus and not turn into a lemon or a carnivorous animal or anything else. The same applies to tigers, lemons, swans and the colour red itself. In our world the existence of concepts is tied up with the stability of the things to which they pertain. We would be unable to make sense of the sorts of things found in Borges' *Tlon* described in the quotation which starts this book. They would be like whatever it was that was seen by the hero of Flann O'Brien's *The Third Policeman* (1974) — literally beyond comprehension and description:

> But what can I say about them? In colour they were not white nor black and certainly bore no intermediate colour; they were far from dark and anything but bright. But strange to say it was not their unprecedented hue that took my attention. They had another quality that made me watch them wild-eyed, dry-throated and with no breathing. I can make no attempt to describe this quality. It took me hours of thought long afterwards to realize why these articles were astonishing. *They lacked an essential property of all known objects.* I cannot call it shape or configuration since shapelesseness is not what I refer to at all. I can only say these objects, not one of which resembled the other, were of no known dimensions. They were not square or rectangular or circular or simply irregularly shaped nor could it be said that their endless variety was due to dimensional dissimilarities. Simply their appearance, if even that word is not in admissible, was not understood by the eye and was in any event indescribable. That is enough to say. (p 117)

If our perceptions were as chaotic as I have described them above, or were they suddenly to become like those belonging to the world of Tlon or the Third Policeman, it would be incorrect to talk of perception of objects at all. To see things as things we need to interact with them and with other members of society through them.[4] This is just as true as its previously mentioned counterpart, that shared perception is required for ordered interaction.

The new riddle of induction

The relationship between concepts and regular expectations has been clearly formulated as a version of the problem of induction by the philosopher, Nelson Goodman. He refers to it as the 'New Riddle of Induction' (Goodman, 1973). In a large part we are actually prepared to get on the bus because we 'induce' that it will not turn tigerish because it has always been bus-ish in the past. Goodman demonstrates the lack of self-evidence in this sort of inference.

He invents a new 'colour'. It is called 'grue', meaning green before some time 't' in the future and blue afterwards. This is a strange sort of colour, and certainly not one that is familiar to us, but it serves a purpose. Take emeralds; every time we have seen an emerald it has been green and this is what makes us think that if we look at one tomorrow it will be green. Now suppose this time 't' were tomorrow morning and suppose that emeralds were in fact 'grue' rather than green. In that case, if we were lucky enough to own an emerald, we would wake up tomorrow morning and find that it was blue since this is what being grue implies. But emeralds are green not grue, so what is the problem?

Why do we take emeralds to be green? Suppose it is because all the emeralds we have ever seen have been green. But since 't' is not until tomorrow morning, what we have seen of emeralds is equally compatible with their being grue. In so far as every emerald I have ever seen has been green, every emerald I have ever seen has also been grue! Grue implies green at every moment in the past. But if every emerald I have ever seen has been grue, then I ought to expect them to continue to be grue. That is, in so far as I think of my emeralds as green, and therefore would be astonished if they are *not green* tomorrow morning, I ought also to think of them as grue, and be flabbergasted if they are not *blue*.

Whether I am astonished or flabbergasted, that is, whether I expect my emeralds to be green or blue tomorrow morning, seems to depend not upon their past qualities but on how I *describe* those qualities. Both green and grue are descriptions that are completely reconcilable with the facts about emeralds known to date, but the term I choose sets up different expectations for the morrow.[5]

There is no need to stick to one new term. For example, 'bleen' means blue today and green tomorrow (see note 5). It is possible to invent as many new terms of this sort as we like, and therefore to choose to be surprised, amazed, dumbstruck, wide-eyed, set back upon our heels, awestruck, and so forth, at any appearance of emeralds whatsoever when we get up in the morning. It all depends upon what we called them in the past. If we decide that past evidence is compatible with their being 'gred' — which it is — then we will be amazed if they are anything other than red. If we call them grellow, then an appearance other than yellow will strike us dumb. But there is no need to stay with colours, nor with emeralds. We might call diamonds 'hoft' meaning hard before 't' and soft afterwards. We might describe buses as 'transivorous', meaning that they have always been means of transport in the past, but we expect them to be carnivorous beasts in the future. In this later case we are back with the problem of consumer confidence and public transport. This world is, once more, like Borges' Tlon where material objects do not exist.[6]

Now as a matter of fact we have none of the above mentioned terms in our language and so we do not encounter the corresponding problems. But the crucial thing is that the terms are not missing because the past appearance of things rules them out of our language; any of the new terms are compatible with the past appearance of all the things that we have ever seen. Goodman's 'solution' to the New Riddle is to say that we see regularity where we have always seen it because of our entrenched linguistic practices. He says that the reason we can 'project' the predicate green — that is base our expectations upon it — and cannot project grue, is that green is the better 'entrenched' predicate. Thus:

> Plainly 'green', as a veteran of earlier and many more projections than 'grue', has the more impressive bibliography. The predicate 'green', we may say, is much better *entrenched* than the predicate 'grue'.
> We are able to draw this distinction only because we start from the record of past actual projections. We could not draw it starting merely from hypotheses and the evidence for them. For every time that 'green' either was actually projected or — so to speak — could have been projected, 'grue' also might have been projected ... (Goodman, 1973, p 94, his stress.)

Thus Goodman's distinction between green and grue is drawn from the comparative history of the terms — grue has no history. There is nothing in nature or logic to forbid the existence of some alternative society where grue was regularly projected and by that fact would be better entrenched than green and therefore properly projectible (see note 5). (Millennial cults — expecting an imminent and massive change in their world — live in a world of grue-like terms!) The lack of 'logical' or natural compulsion is quite clear in another passage of Goodman's:

> I submit that the judgement of projectibility has derived from the habitual projection, rather than the habitual projection from the judgement of projectibility. The reason why only the right predicates happen so luckily to have become well entrenched is just that the well entrenched predicates have thereby become the right ones. (Goodman, 1973, p 98)

or again:

> The suggestion I have been developing here is that such agreement with regularities in what has been observed is a function of our linguistic practices. Thus the line between valid and invalid predictions (or inductions or projections) is drawn upon the basis of how the world is [described and anticipated] and has been described and anticipated in words. (Goodman, 1973, p 121)

This conclusion is important. Our language and our social life are so intermingled that our habits of speech help determine the way we see the world and thus help form the basis for social interaction. However, Goodman's solution still fails to answer the question of how we first came by these particular orderly ways of seeing, how we maintain them and how we develop new ones.[7] Goodman may have solved a philosophical problem but, in spite of the historical flavour of his answer, he has not solved the sociological puzzle.

Most of us are not daily acquainted with emeralds so I will change the example to grass. The trouble with Goodman's explanation of why we see grass as green — namely, that green is an entrenched, projectible, colour — is that grass is also a number of other entrenched, projectible colours. We do not need to invent terms like grue in order to come up with a problem about the colour of grass. If an artist wants to represent grass the chances are that, unless he or she is trying to paint like a child, it will be necessary to use all manner of colours from the pallet as well as green. (Children paint a uniform green strip at the bottom of the picture for the grass and a uniform blue strip for the sky, because they paint what they know to be the case. Thus they paint grass as green, sky as blue, sky as up and grass as down.) To the artist, grass is often brown, frequently yellowish, sometimes blue, sometimes white and sometimes black. It depends on its thickness, its species, the weather, the light and so forth. It is only under very special circumstances that grass is ever green during the night.

Other 'green' things present the same puzzle; to use the language of physics, green things reflect all manner of wavelengths other than those that are taken to correspond to the colour green[8]. It is only under unusual circumstances, such as laboratory experiments with carefully designed specimens and light sources, that a green thing will reflect only the appropriate wavelengths. (It's no good saying that the predominant wavelengths emanating from green things are green wavelengths. There are circumstances in which this just isn't true. For example, it is night for at least half the time when grass certainly isn't recognizable by the green wavelengths emanating from it.)

Since, from time to time, a green thing reflects wavelengths corresponding to all these different colours, and all these other colours are equally entrenched in our language, entrenchment of colour terms does not lead us out of the puzzle. Likewise, since what impinges on our other senses is a jumble of experiences, each individually represented by terms of our language, the existence of entrenched terms does not explain the orderliness of any mode of perception.

It is not difficult to simulate the argument with a little experiment — though it must be born in mind that the exercise is only a metaphor for the philosophical puzzle. Observe a very realistic painting or print or advertisement in a colour magazine. Observe the realistic way in which the objects are represented — each hair, for example, of Holman Hunt's *Scapegoat* or each line of the advertised BMW. Then go closer and closer to the picture. The individual hairs become lost in the smudges and daubs of paint smeared on by the painter with a brush far wider than any represented goat hair; the sharp-edged metal of the BMW turns out to be a blur of different coloured dots.

Wittgenstein and rules

The discussion so far has been in terms of seeing, but the same applies to all perceptual modes, to language and to every cultural activity. Is there perhaps a set of rules for organizing sense experience, fixed within our brains, which we cannot articulate but which we all follow automatically when we see in the same way and when we talk to each other? Such a set of rules, if they could be explicated, would solve the problems of scepticism. It might be argued that, even though we are not conscious of them, it is just such a set of rules that is put into operation in our heads when we see two lots of grass as being the same in their greenness and different to the sky in its blueness. Such a set of rules may be what gives us our *Scapegoats* and our BMW's. But the philosopher Ludwig Wittgenstein's analysis of what it means to follow a rule seems to make such a simple solution untenable.[9]

Seeing the next field grass as belonging to the class of fields of grass, can be thought of as of an example of continuing a regular sequence.

It is a matter of knowing that the next field is a proper continuation of the sequence of fields we have seen in the past. This new field looks like a field, not like something else. It is the 'same' as what went before. What this involves can be explored by looking at a much more straightforward sequence — the numbers '2, 4, 6, 8'. Imagine being asked to continue this sequence in the *same* way. The immediate answer that springs to mind is '10, 12, 14, 16' and, to all intents and purposes, this is indeed the 'correct' answer.

But how do we know it is the correct answer? It cannot be simply a matter of following the rule 'go on in the same way' because this rule allows for a number of possibilities. For example it allows '2, 4, 6, 8, 10, 2, 4, 6, 8, 10, 12, 2, 4, 6, 8, 10, 12, 14' or '2, 4, 6, 8, 2, 4, 6, 8, 2, 4, 6, 8' or '8, 6, 4, 2, 2, 4, 6, 8, 8, 6, 4, 2' or any number of other sequences. For that matter, the instruction 'go on in the same way' could also allow for *who do we appreciate?* as the continuation.

Could it be that the instruction, 'go on in the same way', is not specific enough? Suppose the rule were amended to 'add a 2 and then another 2 and then another and so forth.' But this doesn't fully specify what we are to do either, because that instruction can be followed by writing '82, 822, 822, 8222' or '28, 282, 2282, 22822' or '8²', etc. Each of these amounts to 'adding a 2' in some sense. Now, one knows perfectly well what is meant by 'add 2 to 8': it means, 'come up with 10' even though in a literal sense the other results are not incorrect. However, the question is *how* we know what is meant with such certainty.

One can invent a game to explore the magnitude of this human achievement. 'Awkward Student' shows just how many possible continuations must be ignored in order to continue in the 'right' way. Let one or more persons be the 'Instructor' and one or more be the 'Awkward Student'. The task of the Instructor is to provide a list of instructions to the Student so that given the sequence '2,4,6,8' he or she has to continue it as '10,12,14,16', and so on. The task of the Student is to misunderstand these instructions so that a different continuation is offered such as one of those suggested in the preceding section. The Student, however, must interpret the awkward continuation as a reasonable response to the rules so far provided. As the Instructor fails at each stage he or she (or they) can add to the list of rules, or change them completely in an effort to make the Student go on in the right way. For example, the Instructor's first efforts might be like those offered in the preceding paragraphs. Instead of simply saying 'go on in the same way' the Instructor might try 'add 2 and then another 2 and then another 2' and the Student might make the replies given above. The Instructor might then try to explain what is 'really' meant by 'add' but then the rules will begin to include an indefinitely long sequence covering addition. And so on. The Instructor must not use instructions that rest explicitly upon the taken-for-granted rule-following behaviour in society. Thus the instruction 'do it in the way you normally do it' is outlawed.

The Student, or Students, should always win this game if they are ingenious enough, while the list of rules given by the Instructor, or Instructors, gets longer and longer as tempers fray. Readers do not have to take my word for this, they can try it — but it does take a little practice on the part of the Students. (If the game is tried in the classroom, the lecturer should initially take the difficult role of Awkward Student.)

The game shows that first, rules do not contain the rules for their own application. Second, the notion of 'sameness' is ambiguous. Third, it is not possible to specify fully a rule or 'algorithm'[10] for action in an 'open-system' (where creativity is possible), since if a limited range of responses is not defined in advance then more than one response which satisfies the algorithmical instructions can always be invented.[11] (There is the 'correct' response plus the response which the Awkward Student invents.) Fourth, since in spite of this we all know the correct way to go on, there must be something more to a rule than its specifiability. This 'something' will be explained in a moment, but for now let us call it 'social convention'. The game also shows that evading the normal rules in ways which do not formally break them requires a deal of creative ingenuity — each awkward response is an innovation — which involves inventing a brand new interpretation each time — and that it is easier to do this if there is more than one 'student' to help out; allies are useful even in intellectual battles.

It is important to note that ingenuity is required only because of the necessity of explaining how the chosen continuation 'fits' with the Instructor's latest rule. However, the notion of 'fit' is itself not completely specifiable, and is therefore based on a substratum of social conventions. Thus, even the interaction between the Awkward Student and the Instructor, chaotic though it seems, is based on shared understanding. If there were no shared understanding then the Student would not require any ingenuity at all — a rude noise, or a 'silly answer' would be an adequate response at every stage. Finally, since playing the game is likely to get the Instructor in a very bad temper, it shows how strong the conventions are that bind our normal ways of going on.

Awkward Student will be found to be a useful metaphor for a great deal of what happens at the creative frontiers of 'open systems' such as societies. It certainly leads us to abandon the idea that regularity of perception is a matter of a set of rules within our heads for constraining sense experience into familiar things. There is just too much ambiguity in the notion of the same thing.

The social basis of rules

How do we follow the rules that we do follow, and what does it mean to say that they can be followed because they are social conventions? We have already encountered the social nature of perception with the red bus. We discovered that we couldn't see it as an instance of a bus unless we lived in a society in which other people saw it and used it as a bus — otherwise there would be no one to drive it and no other passengers to give the bus a purpose. Wittgenstein's way of putting this is to say that we know how to go on in the 'same' way because we share a 'form of life'. The rightness of '10, 12, 14, 16' as the continuation of '2, 4, 6, 8' resides in its rightness for everyone sharing our culture. One knows that, in most circumstances, other continuations are mistakes. Others will identify those mistakes even if we forget how to carry on ourselves, or even if we decide to be 'awkward'. Thus, there can be no such thing as a 'private' rule. (Or to put it another way, a private rule would be a vacuous thing, like making a rude noise as a continuation of a number sequence.)

Finally, note that since a rule is not a private thing, but the property of a social group, we might expect rules to vary from group to group. Indeed they do. In other societies, the proper response to the command 'Continue the sequence 2, 4, 6, 8!' might be to make a meal of the instructor. In some other, more familiar, social groups the correct response is in fact 'who do we appreciate?'.

Sociological applications of the philosophical system called 'phenomenology' have provided a language for discussing this cultural variability.[12] Phenomenologists talk of 'taken-for-granted-reality' — a phrase which captures the lack of any sense of accomplishment felt by humans as they order their world. This applies when humans are in the 'natural attitude' — the attitude from which I have been trying to help the reader escape.[13] The content of the natural attitude is different in different groups; that is, their taken-for-granted realities are different. This leads one to expect groups to be able to communicate readily within themselves because of their members' common ways of going on, but equally we would expect difficulty in communication between culturally diverse groups. To use Kuhn's (1962) idiom, the members of different cultures share different 'paradigms', or in Wittgensteinian terms, they live within different forms of life.[14]

(Thomas Berger's *Little Big Man* (1967) provides a graphic inside picture of shifting world views. It carries the reader into the alternate worlds of a frontier American and an Amerindian. The hero is captured and learns to live first as one, then as the other, then as the first again and with each transformation the world seems to make complete sense. The author shows enviable skill in drawing us backwards and forwards between these self-contained, mutually antagonistic and mutually uncomprehending ways of life.)

This leads to a crucial aspect of the approach taken in this book. If cultures differ in their perceptions of the world, then their perceptions and usages cannot be fully explained by reference to what the world is really like. This is the 'relativism' to which I have referred in the Introduction. We must treat our perceptions of the world, for the purposes of this exercise, like 'pictures in the fire'. If the world must be introduced then it should play no more role than the fire in which the pictures are seen. Or better — think of one of those pictures which one constructs by joining numbered dots with pencil lines. Now imagine the world consisting of a large sheet covered in almost infinitesimally small dots. The world is there in the form of the paper but mankind may put the numbers wherever he wishes and in this way can produce any picture. These ideas form the premise of the work discussed in this book and have given rise to a body of work within what has been called the *Empirical Programme of Relativism*, or *EPOR* for short. It rests on the prescription, 'treat descriptive language as though it were about imaginary objects.' I have described the programme as being made up of three stages (Collins, 1981b).[15] This book is meant to provide a widely applicable model of cultural stability and change which also forms a framework for the three stages.

An approach to the problems of inductive inference: joint entrenchment[16]

It has been established that it is only under special circumstances that emeralds reflect only green wavelengths and that the mundane 'appearances' of emeralds neither force us to see them as emeralds nor as green. Nevertheless, we are certain that the correct description is that emeralds are green, and we routinely see whatever colours they reflect as green (unless we decide to attend to them after the fashion of a sophisticated artist). We might say that we know how to follow the correct rules for the description of emeralds and how to continue seeing their colour as green 'as a matter of course'. We are members of a social group which is at one with these rules; we know what misidentifying the colour of an emerald would involve, and we know that in our society others will identify and describe, and indeed see the colour of emeralds in the same way and put us right if we go wrong.

Thus, it is not the greenness of the emeralds that makes them green in spite of the fact that the colour term 'green' is well entrenched in our language. Likewise it is not the 'emeraldness' of emeralds that makes them emeralds. However, it could be that the *joint entrenchment* of the concepts of green and emerald reinforce each other. Thus, if we see a stone, but are not sure what colour it is, and we are told it is an emerald, we are likely to see it as green. Whereas if we see a stone and are not sure what kind it is, but we are told it is green, we are more likely to see it as an emerald than, say, a ruby or a flint. It is not the entrenchment of the concept of 'green' that makes emerald green rather than some other colour but the entrenchment of the notion that emeralds-are-green. To put it another way, greenness and emeraldness are linked within our form of life.

This idea can be extended; there is *multiple entrenchment* of emeralds and their various properties. What is more, each of these properties is itself multiply entrenched with the other objects to which they pertain. What this means is that, in spite of the non-orderliness of the wavelengths reflected by emeralds and other 'green' objects (as can be understood if one strips away the natural attitude), any attempt to bring about a change of usage or perception regarding the colour of a green object such as an emerald would require overturning a whole network of interrelated usages, perceptions and social relations.

The idea of a network of interrelated concepts has been developed by the philosopher, Mary Hesse (eg 1974). It will be discussed and explained further at the beginning of Chapter Six. For the time being, note that the links between concepts in a 'Hesse net' are usually considered to be relations of perceived probability or logical coherence. In my view they are better described as the networks of social institutions that comprise forms of life. I now explain how the network idea serves to stabilize our notions of green and emerald.

The fact that we have the concept green is in part to do with the fact that we have the concept emerald. Our idea of an orderly, unchanging colour called green is tied up, to some extent, with our idea of an orderly unchanging, enduring, set of emeralds. The enduringness of the latter give their colour — green — an enduring quality. The enduring orderly nature of green is however not only a consequence of the enduring quality of emeralds, but is reinforced by its relationship to many other things that are also seen as green. Green is equally reinforced by its being the colour associated with grass, leaves, Ireland, unseasoned wood, the environment, park benches, the field at the centre of villages, and so forth. Likewise, all these things are able to exist for us, in part, because of our stable concept of green. We can account for the stability of green by noting its embeddedness in the whole of our culture; the relationship between greenness and emeralds is just a special example of this. The stability of emeralds, of course, is related not only to the stability of green and other green things, but to the stability of all the other things that share the other qualities of emeralds such as hard things, valuable things, glassy things, desirable things, pretty things, and so forth.

This is how entrenched inductive generalizations are maintained. Their stability is the stability of the forms of life or taken-for-granted practices — ways of going on — in which they are embedded; it is the stability of cultures and their social institutions. What joint entrenchment in networks does not explain is how orderly perceptions are first developed. Nor does it explain how they change. Neither the phenomenological literature, nor the other relevant philosophical literature discusses conceptual change.[17] That old habits of induction die and that new ones develop and gain their own orderliness and stability seems to have passed almost unnoticed. Nearly all the effort has gone into describing the nature of our current taken-for-granted reality.[18]

To explore the problem of change I intend to treat the cases of development of new scientific generalizations which are discussed below as broadly exemplary of all cultural development and innovation.[19]

Science, change and repeatability
Science, like any other cultural activity, rests on a foundation of taken-for-granted reality. Usually scientists spend their time looking at things through the frame of reference that they were given when they were trained. From time to time, however, they try to establish something that does not sit easily within the existing frame of meaning. For example, they might try to establish that some unusual set of data represents a new phenomenon or major discovery. Just as there can be no such thing as a private rule, there can be no such thing as a private discovery. The crucial thing is that others agree that it is a discovery — that they come to act upon it as a matter of course. The discovery, if it is to be a discovery, must precipitate a new set of public rules — a new set of ways of 'going on in the same way'.

Consider the alternative. Suppose I claimed to have discovered that emeralds changed to black at night. I can maintain this as a private conceit for as long as I wish, but it counts for nothing unless it becomes a part of the culture. This might happen — people might start to think of emeralds in the way they think of deciduous trees, as things that change colour at certain times. But if discoveries were private things which did not need public recognition there could be as many discoveries as there are fools.[20]

How do scientists establish that they have made a discovery that should be a new part of the public domain? Press scientists and in the last resort they will defend the validity of their claims by reference to the repeatability of their observations or the replicability of their experiments. This is usually a matter of their *potential* repeatability. Repeatability, or replicability (I will use the terms interchangeably), is the touchstone of common sense philosophy of science. However, as we shall see, the actual replicability of a phenomenon is only a cause of its being seen as replicable in the same way as the colour of emeralds is the cause of their greenness. Rather the belief in the replicability of a new concept or discovery comes hand in hand with the entrenching of the corresponding new elements in the conceptual/institutional network. This network is the fabric of scientific life. Replicability, the vanguard of common sense theories of science, turns out to be as much a philosophical and sociological puzzle as the problem of induction rather than a simple and straightforward test of certain knowledge. It is crucial to separate the simple idea of replicability from the complexities of its practical accomplishment.

The way that common sense theories of science survive in spite of this complexity will become clearer in later chapters. One important and ironic support for the common sense view is that replication of others' findings and results is an activity that is rarely practised! Only in exceptional circumstances is there any reward to be gained from repeating another's work. Science reserves its highest honours for those who do things first, and a confirmation of another's work merely confirms that the other is prizeworthy. A confirmation, if it is to be worth anything in its own right, must be done in an elegant new way or in a manner that will noticeably advance the state of the art. Thus, though scientists will cite replicability as their reason for adhering to belief in discoveries, they are infrequently uncertain enough to need, or to want, to press this idea to its experimental conclusions. For the vast majority of science replicability is an axiom rather than a matter of practice.

But this does not mean that it is not a vital idea. Replicability, in a manner of speaking, is the Supreme Court of the scientific system. In the scientific value system replicability symbolizes the indifference of science to race, creed, class, colour and so forth. It corresponds to what the sociologist Robert Merton (1945) called the 'norm of universality'. Anybody, irrespective of who or what they are, in principle ought to be able to check for themselves through their own experiments that a scientific claim is valid.[21]

It is not only a question of values. Just as with the buses and tigers discussed earlier, the existence and the ordered regularity of natural objects are the same thing. Replication is the scientifically institutionalized counterpart of the stability of perception. It is just that with scientific phenomena one looks through a complex instrument called an experiment. Thus the acceptance of replicability can and should act as a demarcation criterion for objective knowledge. Public agreement to the existence of a new concept implies that its reproducibility can be confidently affirmed even if, as a matter of fact, it is never tested. But use of replicability as a demarcation criterion has to be separated from its use as a test. It is only when the existence of some phenomenon is cast into doubt that attempts are made to use replicability as a test. Otherwise it is a logical counterpart of existence; to affirm one is to affirm the other.

The studies of controversial science discussed in Chapters Four and Five represent instances where there was enough doubt about whether the facts could be replicated to have made it worthwhile for the scientists involved to investigate. What I will establish is that the perceived replicability, or otherwise, which is the conclusion of such tests, really isn't a matter of experiment. I will show that even though the test is made to seem to work, and even though the outcome of the activity surrounding such testing is the demarcation of replicable from non-replicable phenomena, it is not the test of replicability itself which brings about this state of affairs.

Before going on to the case studies I will analyse the idea of replication. This forms the subject of Chapter Two. The analysis is made easier if we distance ourselves from the problem by thinking of science as a purpose built intelligent machine for generating knowledge. It will be found that the difficulties involved in designing this machine are the difficulties involved in adumbrating a rationalist philosophy of science. But first it makes sense to examine briefly the difficulties pertaining to more familiar intelligent machines — intelligent computers.

Rules, induction and artificial intelligence

Observed by only a few commentators, the pigeons of philosophical scepticism and phenomenology are quietly coming home to roost in the nest of artificial intelligence. The next decade or so, which will witness the first large-scale attempts to develop artificially intelligent machines, are going to be of monumental interest to philosophers and philosophically-minded sociologists — optimists and pessimists alike. To see why philosophy and sociology have relevance for the 'fifth generation' of computers I will give a brief example of the problem; we will 'design' a simple artificial intelligence machine — a speech recognizer and transcriber, or automatic secretary.[22] Such a machine must know when the sounds it 'hears' are continuations of sequences of sounds it has heard before. Like other such machines, its principal task will be the recognition of similarity and difference and the application of a set of rules to distinguish and compare things, in this case sounds. Artificial intelligence can be thought of as an interesting experimental application of the ideas in this chapter.

My word-processor recognizes the words I type into it without fail. It can do this because it has to distinguish between only about one hundred different signals — the full range of keys. One might say that it works within a 'closed system' comprising combinations of these one hundred different possibilities. The automatic dictating machine will have to solve problems of meaning recognition far beyond this. Speech recognition is quite a different problem from word processing with keyed-in letters. It requires that a potentially infinite 'open' range of sounds be recognized, not just combinations of a limited range of keys. As we design the speech recognition machine in our imaginations, we will see how it must differ from the word processor.

If we extrapolate into the future so that machine memory and power are far greater than today we could build the machine on the following principles: I speak my whole vocabulary into the machine, word-by-word, and type the corresponding word at the same time. The machine converts each word (phonemes might be more efficient) into an electrical analogue of the sort we would see on the screen of an oscilloscope. Each of these wiggly patterns is converted into digital form and stored in the machine's memory as a sequence of numbers which corresponds to the typed-in word. Given plenty of memory, I might speak each word loudly and softly, quickly and slowly and so forth, so that the machine stores a whole 'envelope' of number sequences or templates corresponding to each word. Thenceforward, when I speak into the machine it will take the sound of each word I utter and match its number pattern against the store of patterns in its memory. When it finds the best fit, it is a simple matter for it to convert the pattern into the corresponding typed letters. Other users of the machine might 'train' it to recognize their voices in the same way. The machine, it should be noted, has to have not only a large memory store, but also the ability to sort through this store at high speed, making many comparisons and selecting the best. (Though exactly what the criteria for 'best' would be is a complicated matter.)

Now this machine ought to be able to recognize the words of the speakers who have 'trained' it, so long as they speak clearly. Let us be generous to the designers and assume that with very clever statistical techniques (working out an algorithm for the 'best fit'), it might even recognize the words of new speakers and speakers who have head-colds. Let it also cope if there is a great deal of background noise or a cross-cutting conversation and even if the speaker does not speak the words clearly and separately but runs them together as we do in normal speech.

However good the statistical algorithms are, this machine will still not be able to untangle fundamentally ambiguous sounds. For example a problem for the word recognizer we have just designed would be to distinguish the endings of these two sentences:

> The sound of a sneeze is '*attishoo*'.
> I'm going to sneeze — pass *a tissue*.

I vocalize these two sentence endings in the same way, but no human being would ever confuse the two in my speech. Here is an example of the *same* sensory stimuli (the same spectrum of sounds, in the language of physics) which are to be properly interpreted under two different rules. In the case of the first sentence interpreting that sound in the 'same way' every time requires that the machine write 'attishoo', whereas in the case of the second sentence interpreting the *same sound* in the *same way* as before requires it to write 'a tissue'.

To disambiguate the two sentence endings the machine seems to need to be able to 'understand' the sentences, not just know how to fit a sound to a template. For example, it needs to understand the concept of 'a sound' so that when it hears 'the sound of . . . is . . .' it will reject 'a tissue' as the correct transcription because it knows that 'a tissue' is not a sound. It will then look only for transcriptions that could represent sounds. Likewise, if the machine knew something about how humans cope with a head cold it could perhaps work out that 'a tissue' would be the sort of thing someone would want prior to a sneeze rather than 'attishoo'. For the purpose of disambiguating this sound in other potential sentence parts, the machine will also need to understand metaphor — 'a tissue of lies' — and nursery rhymes — 'attishoo, attishoo, we all fall down'. Note how much human knowledge is being put directly into the machine to solve this problem. We are now miles away from the word processor or sound recognizer. The machine is beginning to look more and more like *us* with every step. The machine is becoming a social animal. It is acquiring those mysterious abilities that enable us to know when to continue '2, 4, 6, 8' with '10, 12, 14, 16' and when with 'who do we appreciate?'.

I do not see how the machine could ever acquire these abilities without being socialized in the same way as ourselves. It needs to share our form of life so that its joint entrenchments of concepts correspond with those pertaining to our cultural environment. But even if we could imagine ourselves raising a computer in the way some ethologists have raised apes within their households, so that the computer was equipped with the entire social 'script' of the corresponding human child, the 'open system' quality of social life would still not be reproduced. Imagine that this socialized machine is sent out from its nuclear family to work in an office as a secretary. Now imagine that the boss dictates the following joke:

> Two gentlemen from the Indian subcontinent are overheard by a patronizing Englishman as they talk to one another on a bus.
> First Indian: I tell you it's 'woomb' — W-O-O-M-B.
> Second Indian: No, no, no, it's 'whoum' — W-H-O-U-M.
> First Indian: Certainly not Sir, but it could be 'whum' — W-H-U-M.
> Second Indian: No Sir, but perhaps it is 'whoomb' — W-H-O-O-M-B.
> Patronizing Englishman: Excuse me old chaps, couldn't help overhearing. Actually we spell it W-O-M-B — Quite simple really.
> Both Indians: Pardon me Sir, but have you ever heard an Indian elephant break wind?

The point is that not even a human secretary would know how to transcribe this joke — in particular the four spoken versions of the putative elephant sound unless he or she could understand the joke. The only thing that even the most socialized computer would have in its memory store to fit the various sounds would be the word 'womb' and this is what it would have to write in the corresponding spaces. The four words 'woomb', 'whoum', 'whum' and 'whoomb' have been invented especially for the joke. They are brand new. They show what is possible in an open system. They are the equivalent of the creative responses of the Awkward Student. The most remarkable thing is that a human secretary *could* transcribe the joke once he or she had got to the end and understood what the joke required. As I will be arguing, the establishment of novelty is a quintessentially social activity. Who has heard an elephant break wind? The existence of 'woomb', 'whoum', 'whum' and 'whoomb' requires your assent and laughter more than the author's inventiveness or the elephant's flatulence. These four words are, in a sense, about nothing and yet we can all order our lives about them.

Notes

1. For an introduction to the problem of induction and some attempted solutions see Black (1970).

2. Barry Barnes (1976) suggests that our 'natural rationality' is founded in our inductive tendencies.

3. It is tempting to think that we only impute cause to those things that are *always* regularly related and not to those which are fallible. This idea simply does not work. Just think of some of those things about which one is most certain and realize how irregular they are. Take our certainty that the sun will rise tomorrow. As a matter of fact, in Britain, the sun is invisible more frequently than it is visible, even during the day. Why should I believe it is there behind the clouds?

4. This point is made by Peter Winch in his *Idea of a Social Science* (1958).

5. Goodman's 'New Riddle' has given rise to a great deal of critical discussion. A first obvious objection is that 'grue' is not a proper colour term at all because it contains a reference to time whereas other colour terms do not. Actually it is not quite true that our descriptive terms do not contain references to time; for example, 'evergreen' contains a reference to time and so does 'deciduous'.

Goodman's more subtle defence against the time content of grue involves inventing another term, 'bleen', which means the opposite of grue, namely blue before 't' and green afterwards. Now imagine a society in which the terms green and blue were unknown, but the terms grue and bleen were regular features of speech. Imagine they have emeralds in that society which, of course, would be thought of as grue. But imagine those emeralds are in fact like our emeralds so that when the rich citizens of grue-bleen land wake up one morning they find that contrary to expectation their emeralds are green. They would, of course, be flabbergasted, but what would they say to each other? They could not say 'I'm flabbergasted, my emeralds are still green', because they do not have the term green in their vocabularies, or indeed in their conceptual repertoire. What they would say is 'I'm flabbergasted, my emeralds have *changed* from grue to bleen overnight.' In grue-bleen land, it is the term green that requires a reference to time. To describe something as green in grue-bleen land requires that one call it 'grue before 't' and bleen afterwards'! This is what that strange colour 'green' means. Thus, it is not that grue is an odd term because it contains a reference to time; from another point of view, grue does not require a reference to time whereas green does. In this respect too, then, grue and green are symmetrical, just as they are symmetrical in terms of our evidence for assigning either as the colour of emeralds.

6. Jorge Luis Borges in his short story, 'Tlon, Uqbar, Orbis Tertius' (1970) describes Tlon, a land in which there are no material objects. A heretical fable of this world concerns some coins which are lost one day, and found on another. The heretical part of the story is the implication that it is logical to think that the coins have existed in some secret way, hidden from the comprehension of men, during the time that they were lost — that is, unobserved by anyone. Borges points out that the story would be very hard to comprehend in Tlon — a world without enduring objects — because the very notions of 'lost' and 'found' imply continuity of the same objects. The orthodox inhabitants of Tlon ridicule the story in so far as they can understand it. It seems to them rather as though someone were to claim that if a man has a headache one day and then gets better, and another has a headache the next day and gets better, while a third has a headache on a third day, it is the same pain that has passed from head to head! To impute continuity to invisible coins is like imputing existence to invisible pains.

7. For a sophisticated discussion of the intermingling of language and social life see Winch (1958). There is a well known theory called the 'Sapir-Whorf hypothesis' which relates language and conceptual life. The standard example is the Eskimos who, apparently, have many different words for 'snow'. Presumably they see a fascinating, richly detailed, and ever changing patchwork of different things when they look out of their igloos onto what we would perceive as a uniformly white landscape.

Nelson Goodman in his book, *Ways of Worldmaking* (1978), sets out a view which is very similar to the one developed here, at least in terms of its relativism and its treatment of science and art as cultural enterprises which are not obviously epistemologically dissimilar. Both are ways of world making. Rightness and wrongness within worlds, he claims, is a matter of fit with practice embedded in evolving tradition. Outside of such worlds made by humans there is no reality with enough content to be worth 'fighting about'.

8. Experiments on colour vision performed by Dr Land show that it is rather misleading to equate the perceived colour of things with the wavelengths emitted from them. This does not affect my argument, however.

9. It is Wittgenstein's later work to which I refer. The location of many of these ideas is his *Philosophical Investigations* (1953). I take my interpretation of his ideas from Winch (1958). For a very good introduction see Bloor (1983).

10. An algorithm is a finite set of instructions for completing a task. The job of computer programmers is to write algorithms for computers.

11. Bhaskar (1975) erects a whole philosophy upon the distinction between open and closed systems. Unfortunately he draws the dividing line in the wrong place. He accounts for the apparent success of science by its ability to develop theories which work within closed systems; these are experiments conducted within the walls of the laboratory. He believes that the environment of experiments can be controlled so that experimental situations are usefully thought of as closed. As the descriptions of experimental practice to be found in Chapters Three to Six will show, he is wrong in thinking of the experiment as a closed system. The internal workings of a developed computer are probably the nearest thing we have to a closed system in practical terms. This may be because it resembles a theoretical system as near as can be, and these are the only really closed systems.

12. The impact that phenomenology has had on sociology is to turn attention toward the possibility and actuality of different ways of seeing the world. It is the realization of these differences, and their implications for method, that has given rise to the main problems of a scientific (conceived of as positivistic) approach to social 'science'. A two volume history of the phenomenological movement by Spiegelberg (1969) sets out the elements. For the purposes of this book, the clearest and most useful introduction is Roche (1973). Alfred Schutz (1962, 1964) has been most influential in developing the implications of the philosophy for sociological thought. Berger and Luckmann (1967) have produced a widely read treatment of the sociological implications. Berger's little book *Invitation to Sociology* (1963) is an excellent, enjoyable and profound introduction to the implications of phenomenology for social analysis.

13. There is a relationship between phenomenology and the ideas of Wittgenstein. The idea of 'form of life' and the idea of 'taken for granted reality' are close. There are a number of commentaries that draw out the relationship including that of Roche (1973). Other interesting treatments include: Specht (1969), Van Peursen (1959), Munson (1963), Taylor and Ayer (1959) and Gier (1981).

14. Kuhn's ideas are another resource and another way of opening out the problems discussed here. There are many expositions of his work and, in any case, the original (1962) work is readily accessible. For a treatment of the two sides of the debate over the existence of paranormal phenomena as exhibiting different paradigm allegiances see Collins and Pinch (1981 and 1982).

The so-called 'rationality debate' (eg Wilson, 1970) concerned the philosophical problems of understanding strange societies and translating their concepts back into ours; the debate was carried out in terms of the anthropologists' problems. A bizarre feature of the debate was the agreement among many of the writers that in order to learn another language/culture it was necessary to work outward from a few 'bridging concepts' shared between the natives and the stranger. If a few areas of common conceptual ground could be discovered, it was thought that the rest could be assembled on a kind of jig-saw puzzle principle. It was never clear how one could work outwards from bridging concepts if other elements of culture were radically different or, 'incommensurable', to use Kuhn's term. Even odder, however was the setting up of the bridging concept model as the only way of learning a language or conceptual scheme. It seems utterly clear that children do not learn their languages or conceptual schemes in this way since they have no concepts at all when they are newborn which they could use as a bridge between themselves and their parents. Clearly there must be a way of learning which does not involve bridging concepts. Translation is another matter altogether. For some further discussion of the methodology and the anthropologist's problem see Collins (1979, 1983a, 1984a,) and Collins and Pinch (1982).

15. The three stages of EPOR comprise: 1. Demonstrating the interpretative flexibility of experimental data. (The empirically based chapters of this book, as well as

the overall argument which turns around the experimenters' regress, do this adequately, I believe.) 2. Showing the mechanisms by which potentially open-ended debates are actually brought to a close — that is, describing closure mechanisms. (There are a number of closure mechanisms discussed throughout the book and a more systematic schematic treatment is attempted in Chapter Six.) 3. Relating the closure mechanisms to the wider social and political structure. (The way that the core set is linked into the larger social network, as described in Chapter Six, sets out a framework for this task.)

16. The approach offered here differs greatly from the large majority of philosophical treatments of the problem, and should not be confused with them. I do not think that most philosophical approaches resolve the problem at all. Philosophy, *par excellence*, sets problems, which it is much less successful in solving. This is not an anti-philosophical sentiment; the scepticism which philosophy engenders should be a compulsory part of every educational syllabus.

Philosophical approaches have attempted to justify inductive procedures in a number of ways. For example, some have tried to erect supreme principles, such as that the future is always like the past, in order to justify inference in particular cases. The problem then becomes to justify the supreme principles.

One of the most widely favoured attempts at solution depends upon various forms of probabilistic argument. For instance, it can be argued that even though we cannot be certain that the future will be like the past, it is more likely that it will be than that it won't be. The simplest form of this argument implies that seeing a large number of white swans renders the statement 'all swans are white' at least probable if not certain. The trouble is, as Popper has argued, that since the number of swans we have seen, or are ever likely to see, is an infinitesimally small sub-set of the infinite set of swans, however many white swans we see, it renders the statement no more likely to be true. Thus, it should not affect our expectations regarding the next swan we see. However many we have seen we don't know whether we are seeing something that corresponds to a law, or something that corresponds to a run of a single colour on the roulette wheel.

To put this another way, even if our observations of swans could be aggregated and would in fact indicate the long-run tendency of the colour of swans, there is no way we can know this. What is more, we are living in the short-run and we have to make short-run decisions. Thus, since the next set of swans we see might represent a relatively short-run aberration we have no basis for our immediate expectations. This would be true even if we knew that in the very long term the proportion of different coloured swans would tend toward a limit. It is the next swan we are concerned with in this life.

Another probabilistic argument turns on the supposed 'coherence' of our judgements. This approach takes betting behaviour as a metaphor. It is suggested that the 'rational' gambler will never make bets where the outcome is that he or she is bound to lose. The arrangement of bets, which the rational better would *not* accept, is known as a 'Dutch Book'. If it is assumed that rational actors avoid a Dutch Book then something can be said about the disposition of their beliefs about the future. That is, they must be interrelated in such a way as to avoid certain 'loss' if the beliefs were somehow aggregated.

This does not get one a great deal further however. As Hesse (1974) points out, if it works, it says nothing about any *particular* initial belief, it only puts limits on the way beliefs can be interrelated; it would allow us, therefore, to start out with some ridiculously counterinductive dispositions. It would, for example, not prevent us from giving odds of one thousand to one on a particular outcome of a coin tossing experiment that we knew to be fair so long as the odds we gave on the other outcome corresponded. In fact we bet upon far more 'inductive' information than this.

In any case, humans just do not make their decisions in particular instances by reference to their preferences on every other interrelated case. The relations of the items in our conceptual network to one another are so complex as to make this impossible. Even if such complexities could be grasped and agreed, and even if they could form the basis for a complex calculation of odds, they would still not circumvent the conventional foundation of our inductive propensities. There may be more adequate probabilistic approaches (see Hesse, 1974, for a discussion), but I do not believe they are likely to reflect our real life inductive behaviour.

An interesting sidelight on this point is provided by the experiences of researchers in the field of computer 'expert systems' or knowledge-based engineering. Some models of expert decision making rest on such a structure of probabilities, but 'knowledge engineers' find it almost impossible to persuade human experts to provide them with the simple probabilities of various happenings that they require to stock their information base. Humans just do not think naturally in probabilistic terms.

17. Thomas Kuhn (1962) is a rare exception in trying to describe processes of change, and revolutions in particular.

18. The difficulty in explaining change within these philosophies seems to arise out of their explanation of human action in terms of conceptual life. The causal arrow runs all one way, as it were. Winch's (1958) book has been rightly criticized in this respect — for example by Gellner (1974) and Bloor (1973).

What Winch has done is to use Wittgenstein's ideas to argue the identity of social and conceptual life — that social relations are embodied logical relations while logical relations are the abstract counterpart of social relations. This, however, gives one just as much right to argue from changes in social life to changes in conceptual life. Actually the two go hand in hand. Where the argument runs entirely from conceptual to social there is no motor of change to be found. Why should a stable system of concepts ever change? A great deal of Winch's argument is devoted to showing how even what appear to us to be ridiculous parts of others' conceptual systems can be quite easily fitted into a consistent framework without any appearance of 'irrationality'. Thus, conceptual frameworks will never generate trouble spontaneously; contradictions will not become apparent by themselves. We must take it that people want change for reasons which emanate outside a closed conceptual system, if change is to be understood.

Kuhn is the only recent philosopher of science to try to cope seriously with conceptual change, but his efforts are not entirely successful because he places too little emphasis on the social determinants of scientific revolutions. This is clear in the ambiguity of the notion of 'anomaly' in his writing. On the one hand, changes seem to be brought about, at least in part, by a build up of anomalies — these presumably must irritate scientists. But on the other hand, calling something an anomaly is a device for ignoring it — it enables it to be swept under the carpet, as it were, so that the appearance of the coherence of the paradigm is maintained. We have, then, an unsatisfactory idea of the build-up of an irritating mass of soothing things as a partial condition of the occurrence of a scientific revolution. Clearly, something more is needed. When and why do the soothing things become irritating? The notion of anomaly cannot, of itself, explain this.

19. See methodological appendix for the relationship of these ideas to some other contemporary, and closely related, writers.

20. Brannigan's book *The Social Basis of Scientific Discoveries* (1981) makes the point that the crucial feature of a discovery is the assignment of the label 'discovery' rather than any set of activities or observations. As he points out, the same set of observations and activities, carried out a second or third time, does not constitute a discovery for anyone except the individual.

21. In this sense, the sociologist Harriet Zuckerman (1977) is quite right in her claim

that replicability is the foundation stone of the value system of science. But it must be borne in mind that it is essentially a theoretical idea, and that the test of replicability cannot be a way of divining frauds, cheats and fools in the way she thinks. The sheer infrequency of replication shows this, but the more profound point is the circular nature of replication; this will be explained in later chapters.

22. The example of a speech recognition machine is far simpler than, say, a language translating machine, which would have to cope with another more subtle set of ambiguities. As will be seen, however, all such tasks seem to require that the machine share in human culture if it is to mimic human abilities. All this is explained in a book called *What Computers Can't Do* by Hubert Dreyfus (1979). In spite of its 'popular' title, Dreyfus's arguments are very sophisticated and rooted in phenomenological and Wittgensteinian ideas. He argues that if computers are to move from their current situation of limited symbolic environments into the infinite real world the programmers must solve the problems set by Edmund Husserl's phenomenological programme in philosophy — a task with which two thousand years of philosophical effort (under different names) has failed to make much progress.

There are criticisms of Dreyfus's argument. In the main these seem to take the form that whatever he has said cannot be done turns out to have been done a few years later — and then he says that this was not really a proper test. Most critics do not seem to understand nor to tackle the underlying philosophical point. For criticisms, see McCorduck (1979) and Boden (1977).

Chapter Two

The Idea of Replication

Of mice and men:
terrestrial science as a research machine
In a book called *The Hitch-Hiker's Guide to the Galaxy* (Adams, 1979), the Earth turns out to have been commissioned by mice so that they could use it as a large computer. In this chapter I will adopt the perspective of these mice to view human science as a computer-like research machine. In particular I want to look at the murine approach to the testing of facts by experiment and replication. Let us start by imagining 'Poppa mouse' explaining the 'replication algorithm' as earthlings would live it. Actually, we need do no more than take a couple of quotations from our own terrestrial philosopher, Sir Karl Popper. The following might well be lines belonging to that very replication programme as embodied in the human mind:

> Only when certain events recur in accordance with rules or regularities, as in the case of repeatable experiments, can our observations be tested — in principle — by anyone. We do not take even our own observations quite seriously, or accept them as scientific observations, until we have repeated and tested them. Only by such repetitions can we convince ourselves that we are not dealing with a mere isolated 'coincidence', but with events which, on account of their regularity and reproducibility, are in principle inter-subjectively testable.

and

> Any empirical scientific statement can be presented (by describing experimental arrangements, etc.) in such a way that anyone who has learned the relevant technique can test it. (Popper 1959, pp 45 and 99)

If Popper's thoughts really are part of an algorithm there is a bug in the programme. This is because he has also written things which seem to make the above instructions difficult to follow. Elsewhere he has said:

> All repetitions which we experience are *approximate repetitions*; and by saying that a replication is approximate I mean that the repetition B of an event A is not identical with A, or indistinguishable from A, but only *more or less similar* to A. But if repetition is thus based upon mere similarity, it must share one of the main characterisitics of similarity; that is its relativity. Two things which are similar are always similar *in certain respects*... The remark may be added that for any given finite group or set of things, however variously they may be chosen, we can, with a little

ingenuity, find always points of view such that all the things belonging to that set are similar (or partially equal) if considered from one of these points of view; which means that anything can be said to be a repetition of anything, if only we adopt the appropriate point of view. This shows how naive it is to look upon repetition as something ultimate or given. (Popper 1959, pp 420, 422)

Popper, of course, knows about the problem of inductive inference, but he has not drawn the connection between it and the matter of ascribing similarity and difference to experimental procedures and their outcomes. This is crucial.[1]

The bug would show itself when the mouse computer failed to complete its tasks in a straightforward way. Some results would seem to come in much more slowly than they should and others would never appear. It would turn out that the 'Popper problem' was echoed throughout the working parts of the machine. Chapters Four and Five and the following examples illustrate the difficulties.

Psychology

The science of psychology exhibits the Popper problem in almost ideal form. Terrestrial psychologists — merely components in a machine so far as the mice are concerned — started to do experiments with admirable, perhaps even excessive, zeal in the early part of the twentieth century. They used Poppa mouse's replication algorithm. Thus, in 1926 a psychologist called Dennis wrote:

> Proof in science is merely repeatability . . . what has occurred once under given conditions will occur again if the same conditions are established . . . The only question concerns the accuracy and completeness of the statement of conditions. . . . proof is not begun until the conditions of the experiment, as well as the result, are so accurately described that another person, from description alone, can repeat the experiment (quoted in Friedman, 1976).

However, in 1967 another psychologist (Friedman), in large part inspired by the work of Rosenthal (1966), wrote that the recipe for the exact repetition of a psychological experiment could not be transmitted since the crucial variables that would need to be controlled had not yet been delineated. Rosenthal had revealed the existence of 'experimenter expectancy effects' in psychological experimentation. This meant that the results of experiments tended to come out in a way that favoured the experimenter's expectation of how they ought to come out however much he or she tried to avoid bias.[2]

If expectancy effects are real then a crucial variable in a psychological experiment is the experimenter's prior *beliefs* about what ought to be the correct results. That means that a positive replication would not be convincing if it were done by someone who thought the result ought to be positive since they would be likely to have biased the work in a positive direction. On the other hand, it could be argued that a replication where the second experimenter *did not* share the first experimenter's views was insufficiently *like* the first experiment to count as an exact replication! Thus, Dennis's prescriptions begged the question of the meaning of 'same conditions' and 'another person'. The quality of the other person, that is his or her prior beliefs, seemed to be a part of the conditions. Thus, while Dennis's statement coincides with the replication algorithm embodied within the first two quotations taken from Popper, arguments about the significance of expectancy effects exemplify the 'bug' engendered by the strictures on similarity and difference expressed in the third quotation.

Experimenter expectancy effects were merely one outcome of what Friedman treated as the special problems of psychological experimentation. He pointed out that where subjects are human the experiment is a social interaction so that even miniscule, or subliminal, variations in timing, cueing, presentation of self, eye contact, manner of speaking and so forth, can affect the responses of experimental subjects. He felt that as long as this special problem of psychology was not recognized, experimenters could, in effect, cheat:

> The point is that in contemporary psychology, experimenters can vary rooms, times, days, seasons, sexes, regions, experimenters, tables and chairs and still be engaged in replicating, so far as their colleagues are concerned, the same experiment ... The rule is in essence: do not vary anything which makes a difference in the subject's responses; that is what it means to do the same experiment. (Friedman, 1976, p 149)

Thus psychology, after a promising start as a sub-system of 'Terraputer', turned in on itself. Even now, however, it treated itself as a special case rather than seeing that it merely exemplified a more widespread syndrome.

Parapsychology
Another notoriously slow running area was parapsychology. A hundred years of effort had failed to give rise to consensus about the existence of phenomena such as telepathy, clairvoyance, mind-over-matter (psychokinesis) and so forth. Parapsychology was afflicted by the same syndrome. Thus whereas some parapsychologists were quite certain that it was repeatibility that counted, others were unsure what this meant. In 1956 the following discussion took place among a group of parapsychologists (Wolstenholme and Millar, 1956). It illustrates the widespread confusion about the meaning of replication which arises when close and detailed consideration is given to the meaning of the concept.[3]

West: [the best experiments in parapsychology] fall short of the requirements for universal scientific conviction for several reasons, the chief one being that they are more in the nature of demonstrations than repeatable experiments ... No demonstration, however well done can take the place of an experiment that can be repeated by anyone who cares to make the effort. (p 17)

Nicol: ... repeatable experimentation [means] ... the designing of an experimental set up which, if found in practice to produce a significant effect, can be repeated by any competent person at any time in the foreseeable future with approximately similar significant results. After thirty years, psychical researchers have failed to produce one repeatable experiment. (p 28)

Gaddum: One kind of evidence that I feel is really convincing is some people can do it and some cannot. I mean, if it was the sort of thing that anybody might do sometimes, I should not believe in it so readily as in the fact that once you have got a man like Shackleton [an apparently successful psychic subject], you can apparently make repeatable experiments with him. I do not understand why it is suggested that it is not a repeatable experiment, if day after day you can always get significant scores. (p 39)

Pratt: I cannot see how there is any serious question to be raised regarding the repeatability of an experiment when Dr Soal and Mrs Goldney, working with their selected subjects week after week, were able to get results and were able to do so with visitors brought in to witness and even to take charge of their experiments ... to his way of thinking he had successfully repeated the Duke work [an apparently successful series of experiments at Duke University]. The thing to keep in mind, it seems to me, is the *essential* nature or the *essential* feature of the experiment, and if similar results are obtained by a number of people working in different places, then I think that in every real meaning of the term the results are repeatable. (Pratt's stress, p 40)

Wasserman: ... I do not think we can use the word repeatable without caution. ... there are two types of repeatability ... repeatability within a single experiment ... and between different experiments, and these two types of repeatability are entirely different things.

Let me also discuss ... repeatability *at will* ... Looking for such a type of repeatability is looking for a Mare's nest. Take for instance experiments in cosmic physics. Sometimes you find a meson track, sometimes you do not. You can set out with iron determination to find a track and come home empty handed. Nevertheless, physicists agree that these tracks even if rarely found are important. In other words, if we have a rare event we cannot expect it to be repeatable at will. We must simply distinguish between high and low inductive probabilities, and this I think Mr Nicol does not seem to appreciate. (Wasserman's stress, p 41)

Langdon-Davies: It seems to me ... rather interesting that the parapsychologists all fall over backwards and say 'We must have repeatability', but that the biologists at this symposium say that this does not impress them as a primary necessity. (p 42)

Spencer-Brown: As someone pointed out, we can get repeatability within a single experiment, but this is not what is meant by repeatability in science. We want results which are not only consistent in one experiment, but which can be observed to recur in further experiments. This does not necessarily mean that they must be repeatable at will. A total eclipse of the Sun is not repeatable at will: nevertheless it is demonstrably repeatable — we can give the recipe for its repetition. And this is the minimum we look for in science. We must be able to give a recipe. (p 44)

West: ... some experimenters may be incapable of repeating the results which other experimenters are able to obtain. That is what I meant by factors of repeatability. (p 45)

Nicol: ... Dr Wasserman is clearly giving the word a meaning and definition that are not usually attached to it in scientific thought. Dr West apparently, and certainly I, were unaware that the repeatability problem, which has defeated many years of effort, has in fact been solved and we did not notice it.

... [The Shackleton case and so forth] ... do not constitute repetition as I understand it. I mean by repetition — and I am rather under the impression that in the physical and other sciences this is also the understanding — that you design an experiment which any competent person can repeat with approximately the same results, and now come the essential words, *at any time in the foreseeable future.*

If ... you can get a result repeated over and over again, then you have only to describe it to any other competent person, psychologist, physicist, medical man or anyone else, invite him to do it for himself, and he will get the same results. When you can do that successfully you will have won half the battle for recognition. (Nicol's stress, p 48)

Thus in parapsychology too, the easy demarcation of the true from the false by reference to the criterion of replicability failed in the face of doubts about the meaning of the term for practical purposes. To this day a verdict on the phenomena studied by parapsychology has not been delivered.

Philosopher-mice and an analytical theory of replication[4]

Let us now imagine that philosopher-mice tried to reconstruct the rules which humans were supposed to follow. They might feel that earthlings had been given too much leeway. The programme had grown beyond comprehension. Thus, it might well be that some humans were 'biased' or acting 'irrationally'. Though humans did seem to work in concert most of the time the rules of synchronized action were difficult to specify. It was when they failed to work concertedly that the machine became sluggish. If the hidden rules of proper concerted action in the domain of science could be rediscovered then it would be possible to make certain that disagreements would never arise in the first place, or at least, be rapidly settled; every human component could be made to see things in a way which corresponded exactly with his or her position in the logical scheme. In short, we are imagining that the philosopher-mice wanted to develop what terrestrial philosophers call a prescriptive philosophy of science.

Let us imagine that the mice decided to develop an analytic theory of replication. They had studied enough of the curious problems in the machine to know that simple solutions, such as those embodied in Dennis and Popper (in his more carefree moments), could never work. Somehow the complexities had to be taken into account.

The analytic theory

For an experiment to be a test of a previous result it must be neither exactly the same nor too different. Take a pair of experiments — one that gives rise to a new result and a subsequent test. If the second experiment is too like the first then it will not add any confirmatory information. The extreme case where every aspect of the second experiment is literally identical to the first is not even a separate experiment. Under these circumstances the second experiment would amount to no more than reading the first experimental report for a second time.

Confirmatory power, then, seems to increase as the difference between a confirming experiment and the initial experiment increases. For example, imagine that there is just a very small difference; it might be a small difference in the time when the two sets of experimental results were generated, such as with an almost immediate second observation of the result by the same observer. We might describe this second observation as belonging to the same experimental 'run' or it may be just that a meter reading, or some other observational outcome, remains steady for more than a split second.

A period of observational steadiness of this sort certainly does confirm the 'first impression' given by the initial glimpse of the result. Each subsequent reading, or each subsequent moment during which the meter stays steady, confirms the initial impression further. A whole new run the next day provides much more confirming evidence. A run conducted by another experimenter on the same apparatus is still more impressive, and a confirming run observed with a similar apparatus built and run by another experimenter is even better. Still more convincing is the same result generated by an apparatus designed upon different principles, for then it is certain that the result is not simply an artifact of the particular equipment or particular design of the original set up. If this final demonstration were to come as a surprise to an experimenter who initially believed the opposite outcome the more likely, then the confirming power is greater still.[5]

However, this conclusion — that the more different an experiment to its predecessor, the more confirming power it has — provides only one side of an epistemological tug-of-war. The other side is best understood by taking another extreme example. Suppose some startling new result has been produced in, say, physics. Imagine that it has been subsequently confirmed by someone quite different in background, who did not initially believe that the first result was correct, and who used apparatus that was very different to the original in concept, design and theoretical premise. Should this be a cause for celebration on the part of the first experimenter? The answer must be 'yes' if the reasoning of the previous paragraph is the whole of the story, but suppose the second experimenter were a sceptical fairground gypsy who had generated the confirmatory result by reading the entrails of a goat! Even though the differences between first experiment and second experiment were maximized, the first experimenter would not be pleased. Indeed, if the entrail-results were to be cited as supporting evidence, the effect would probably be damaging.

If we move back in stages toward lesser degrees of difference we can now see that the situation steadily improves, just as it did as we moved toward greater degrees of difference in the first part of the argument. Thus, if the gypsy had used some old technical equipment rather than goat-entrails, it would have looked a little better. If the gypsy is replaced by a high school student it looks still better (though in most circumstances best not reported).[6] If the high school student had used good apparatus, then things would be better, and likewise if it were a first-rate physicist who had used the poor apparatus. Thus working backward from the extremes of difference we get more and more confirming power. It looks as though the optimum point will be somewhere toward the middle.

Unfortunately, the theoretical tug-of-war does not have a stable equilibrium point. What is treated as the best solution varies as a function of a variety of factors. For example, the less that is known about an area the more power a very similar positive experiment has to confirm the initial result. This is because, in the absence of a well-worked-out set of crucial variables, any change in the experimental situation, however trivial in appearance, may well entail invisible but significant changes in conditions. In a poorly understood area, scientists just do not know enough to be able to guarantee that an experiment which looks just the same as another is the same in essence. They are unable to guarantee that its results will be the same and thus observation of the results does add confirmation — it amounts to more than reading the same experimental report a second time. As more becomes known about an area however, the confirmatory power of similar-looking experiments becomes less. This is why the experiments performed every day in schools and universities as part of the scientific training of students have no confirming power; in no way are they *tests* of the results they are supposed to reveal.

Another complicating factor is that, though confirming power usually increases as experiments differ more (apart from the extreme case), there are circumstances in which power increases with similarity all the way to the extreme of near identity of the second experiment with the first. These circumstances arise when the second experiment is intended to disconfirm the first. This is because if a second experiment fails to see a claimed result, differences of design between first and second may be invoked as the cause of the failure; it will be said that the second experiment has not been done according to the instructions. Thus the strength of a disconfirmation goes up as the second experiment approaches identity with the first. The extreme case is when an experimenter's own 'first impression' is not confirmed by his own second look at his meter, or whatever.

Thus the attempt to produce a theory of replication has only produced a bit of a mess — not the sort of thing that would help the philosopher-mice in their quest for a prescriptive theory. To turn it into something more orderly it is possible to make an adjustment reminiscent of terrestrial philosopher Imre Lakatos.[7] The adjustment involves dividing up the principles upon which an experiment might be designed into two sets: the set which is an acceptable part of science and the set which is not. Likewise, *experimenters* can be divided up into *bona fide* scientists and unqualified upstarts or pseudo-scientists. It is not hard for us to judge on which side of the dividing line the gypsy and the goat entrails fall. Given this division we can say that confirmatory power of experiments increases with diversity of design and personnel within legitimate science, but begins to decrease when the diversity becomes so great as to move into the pseudo-science area. Finally it reaches an area of negative confirmatory power for something as extreme as the gypsy. For disconfirmatory experiments, power decreases with degree of diversity within legitimate science while disconfirmatory attempts within pseudo-science are of no measurable value. Figures 1a and b represent the cases of confirmation and disconfirmation.

Figure 1a represents the case of a new claim in an ill-understood area. There is considerable confirmatory power to be had from an experiment only slightly to the right of the origin. Only the identical experiment gives no confirmation at all; the origin represents this point. In figure 1b the graph tends towards infinity at zero difference. This makes sense because at the infinite point the phenomenon would never have been seen in the first place.[8]

Here then is a theory, complete with diagrams, which might be used to organize the work of scientists on Terraputer. Undoubtedly it captures a lot of what we know to be terrestrial reality. Scientists certainly think more of an experimental confirmation done by a distant enemy of the original claimant than one done by a colleague or

Figure 1a. Experimental confirmation

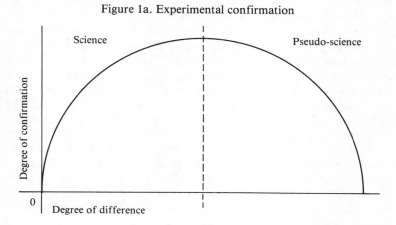

Figure 1b. Experimental disconfirmation

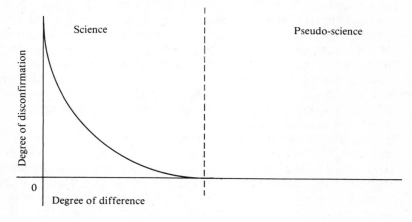

friend. If it is a disconfirmation, however, it will seen as more impressive coming from a (short-lived!) friend.[9]

Nevertheless, the theory contains awesome gaps. Reading back over the arguments of the parapsychologists presented earlier in the chapter it is clear that the theory has not dealt with questions such as the ease of repeatability (for example, in the sense of 'repeatability at will') and has not dealt with failed attempts at replication (whether, or when, these are to be counted as disconfirmations). A more glaring gap, which should be irritating the reader of this book, is the question of uniformity of perception of the degree of difference and similarity between one experiment and another. This is where the difficulty first arose with Popper. It is all very well to develop an abstract theory of this type, but where humans disagree about which experiment was like which other, or disagree about which experimenter is really a pseudo-scientist, how are the philosopher-mice to settle matters without actually doing the science themselves?

To understand the problems the mice need a different starting point which assumes less about the transparency of judgments of similarity and difference.

An empirical model of replication
Instead of starting with abstract notions of similarity and difference, let us suppose that the mice explore the problems of these concepts as they present themselves to humans. Suppose they agree to take some result 'r', first generated at time 't_1', and examine every subsequent activity that takes place on Earth up to, say, time 't_2'. The idea is to find out how terrestrial scientists decide if some sub-set of all the multifarious activities comprise replication of 'r'. If it is possible to discover how this decision is made in practice, then they would effectively discover the rules of replication. The programme bug could then be eradicated.

The mice would then be faced with an enormous problem of sorting. They would have to take an almost infinite number of terrestrial activities, ranging from the noble and the cerebral through to the trivial and the disgusting, and narrow them down to just that set which comprised replications of 'r'. Suppose they agree to try to do this by eliminating activities through a number of sorting stages. The exact number and nature of the stages would be arbitrary to some extent, but any sensible scheme would filter down from general to more specific sorting criteria. The following is the type of schema they might adopt.[10]

A sorting schema for the determination of 'replicatedness'
Level one: Eliminate all activities not to do with the subject of 'r'.
Level two: Eliminate all activities that are not scientific.
Level three: Eliminate all activities where the identity of the experimenter is inappropriate.
Level four: Eliminate all activities that are not experiments.
Level five: Eliminate all experiments that are not competent copies of the original.
Level six: Divide the remainder into those which are positive and those which are negative.
Level seven: Decide whether 'r' has been replicated.

By working through this schema an inventory of questions concerning the nature of replication can be developed. We will discover that at each level we are unable to provide clear cut demarcation criteria which allow us to move on. Fortunately, there is no need to be held up by this problem for we are merely following the progress and thought processes of the mice. Where we find an unsolved problem we will simply forget about it for the time being. The mice, we will assume, are able to solve the demarcation problems at each stage by invoking *murine rules*

Level One: In general this stage does not present too much trouble, though there are some areas of endeavour where problems arise; once again, parapsychology is a good example. It has been argued that all deliberate bodily movement exemplifies psychokinesis since it involves the control of the matter of the body by the mind. At first glance, the intimate relationship between mind and body in this instance seems to make the comparison with standard psychokinetic feats rather strained. Nevertheless, it can be argued that the only difference is that in the case of bodily movement the 'gap' between mind and object moved is hidden somewhere inside the body rather than being obvious for all to see. Without going into detail we may imagine that demarcating between activities that are and are not to do with the phenomenon of psychokinesis would engender a fair amount of squeaking among the mice. There are other areas of science where the matter is equally unclear and humans find it difficult to settle their differences. We will, however, accept that there is a murine rule that can settle the matter and move on.

Level Two: At this stage it ought to be possible to eliminate troublesome pseudo-science. Anything like a goat-liver divination of 'r' would be eliminated even though it was an activity to do with the subject of 'r' and therefore would have passed through the previous level of sorting. The demarcation is not entirely without its problems; disagreements about this sort of thing are endemic among human scientists. What is more, so long as terrestrial science is developing there are bound to be activities which were once thought to be 'scientific' and are no longer thought to be so — alchemy comes to mind — and there will be others once thought to be beyond the pale of science proper but subsequently brought within — acupuncture is an example. Nevertheless, let us pass on to the next stage on the assumption that the mice have a way to deal with this problem.

Level Three: Here pseudo-scientists can be eliminated even though they appear to be going through the motions of doing proper scientific work. Criteria concerning appropriate background, training and personal qualities would need to be developed; and trouble can be anticipated once more in the light of the developing nature of science, and in the light of the discussion in the previous section on the analytic theory.

At this level the mice will have to decide on what will count as suitable social and cognitive relationships between experimenter and replicator. Think how little confirmatory power is accorded to replications carried out by scientists' mothers! Clearly such a social relationship is outside the proper boundaries. Likewise, replications by other blood relations, or by colleagues who work in the same laboratory, ought perhaps to be given a low value. The difficulty is that the case of the relationship of identity — self-replication — is sometimes valued. For an example of this, see the arguments by Pratt in the extracts from parapsychologists' discussion given above. Also, the point that any replication has value in an ill-understood area, as explained in the last section, is relevant.

In the section on psychology, expectancy effects were discussed. This brings out the question of the permissible range of beliefs of a proper replicator. In parapsychology, for example, there has been a real difficulty since the discovery of the so-called 'sheep-goat' effect (Schmeidler, 1958). This is the claim that subjects and experimenters who are more sympathetic to parapsychological ideas (the sheep) are more readily able to manifest, or create, the conditions for the production of paranormal phenomena. Sceptics (goats) on the other hand inhibit the appearance of effects. If the sheep-goat effect is accepted, it means that it is only those with similar expectancies (positive beliefs) who are able to replicate parapsychological experiments. Thus, while the theory of expectancy effects in psychology might lead one to disqualify replicators with similar 'biases' to the initial claimants, the sheep-goat effect would disqualify those who had different 'prejudices'.[11]

In any case, as will be seen, positive replications by critics are exceptionally rare events in science. If the mice were to introduce a rule that restricted proper replicators to the class of critics or the socially estranged, then the resulting programme would run very slowly indeed.

By contrast, note amongst the parapsychologists' comments, that

Nicol thinks that any competent person in the foreseeable future ought to be able to produce a result and this includes psychologists, medical men, physicists or 'anyone else'. Let us assume there is a murine solution and pass on.

Level Four: What is an experiment? Scientists very rarely form their initial beliefs through doing experiments. Most experiments are so difficult and time-consuming that it would be foolish to begin unless one had a firm idea that the result would be useful. Nevertheless, experiments have to be done to convince others, or perhaps to 'certify' a finding for those who are ready to be convinced.[12]

As we shall see when we turn in later chapters to the reports of actual experimental work, experiments hardly ever work the first time; indeed, they hardly ever work at all. Thus, any sensible experimenter ought to expect that most of what he or she does in the way of practical activity will be trial and (mostly) error. It will comprise not proper experiments, but one *preliminary run* after another. This presents serious problems for a statistical science because the classification of a piece of practical activity as an 'experiment', as opposed to a preliminary run, or practice trial, makes a difference to the aggregate statistics.

Again, the case of parapsychology is instructive. In card guessing experiments, a 'subject' tries to guess the symbols on target cards without being able to see them. Critics have suggested that the mass of successful experiments of this sort (where statistical analysis shows that the subject has succeeded in guessing a significant but only slightly greater number of targets than ought to be the case according to chance) might well be offset by the huge number of experiments that do not meet with success and are therefore never reported. The reports of negative experiments, it is suggested, remain in experimenters' filing cabinets never to see the light of publication. This is known as the 'file drawer problem'. Since statistical theory only predicts the outcome of long runs of trials, the existence of a small number of successful trials, offset by a large number of unsuccessful ones, would say nothing about the causes of the success. Some success would be expected purely by chance.[13] Thus the classification of a piece of work as an experiment or preliminary run is vital in parapsychology and in other statistical sciences.

It seems to me that since most experiments are delicate, and fail to work most of the time (see next chapters for an empirically based discussion of this point), then they are not properly reportable as experiments, even in a statistical science, until an appropriate level of skill has been attained by the experimenter. But what this would mean is that no-one ought to report a negative result until they had demonstrated their skill by producing a positive result. This must mean that all experimenter-critics would be disqualified at the outset; all their results should be viewed as preliminary runs. The strongest form of the argument would apply equally to statistical and non-statistical sciences. Attractive as this is, in so far as it would disbar 'pundits' from commenting on others' experimental work, the argument as a whole looks like a *reductio ad absurdum*.

We have then, something of an impasse, the practical significance of which will be seen in later chapters. For the moment let us note the demand for 'replicability at will' expressed by Nicol in the parapsychologists' quotations and the distinction made by West between a 'demonstration' and a repeatable experiment, and leave the mice to it.

Level Five: It is at this level that nearly all terrestrial efforts to produce a 'rational' theory of replication, or a statistical calculus for aggregating negative and positive results, break down. The problems are brought out by examining Rosenthal's (1978) attempt to settle a dispute through the use of such a calculus.

Rosenthal discussed no less than 345 studies of his own expectancy effect hypothesis.[14] He and his colleagues developed a statistical technique for aggregating the positive and the negative, or null, results. Of the 345 results, roughly two thirds were negative or null, and roughly one third were positive. Nevertheless, when the calculation was performed, it turned out that the positive statistics far outweighed the negative ones so that the hypothesis was clearly upheld.

This sort of calculation seems perfectly reasonable so long as it is assumed that the variation among experiments is entirely accounted for by random fluctuations in unknown background variables. Under these circumstances it is only to be expected that some of the results would turn out to be negative or null and it would be quite reasonable to argue in this statistical fashion (see for example, note 13). However, it might be suggested that since experiments are such delicate things, it would only be reasonable to expect that *most* of the 345 experiments discussed were defective in one way or another, and it would not therefore be unreasonable to suppose that all, or most, of the one third positive experiments were defective. The statistical calculus cannot distinguish between well done and badly done experiments and, of course, the most elaborate and immaculate calculation is useless if it is applied to defective work.

Thus, attempts to generate an algorithm for the aggregation of experimental results turn on the possibility of distinguishing between competent and incompetent experiments. As a matter of fact, Rosenthal recognizes this in his own (1978) article where he suggests that the statistics still come out in favour of the hypothesis when attempts are made to introduce criteria of excellence. To establish these he looks for special methodological safeguards or for special scrutiny of experimental procedure (as with work done toward a PhD).[15]

It is important to note that I am using the term 'incompetence' in a special way. The term is not meant to imply cognitive or manipulative incompetence though these would be included. By an incompetent experiment I mean one in which the results do not support the hypothesis in the way that is claimed. Thus an experiment can be incompetently performed, in this sense, even if the data have been generated properly but some alternative explanation for them has been overlooked. This does not imply 'manipulative incompetence'. The question of perceptions of competence and incompetence is to be discussed at length in the remainder of the book. For the time being let us note that among the parapsychologists Spencer-Brown, echoing the psychologist Dennis, believes that it is possible to give a recipe for the repetition of an experiment and that Pratt thinks that only the essential nature of the experiment needs repeating. We must assume that the mice can find their own extra-terrestrial solution.

Level Six: It might be thought that there can be little difficulty in deciding which experiments have produced positive and which negative results. However, there remains a number of problems. The major one which will be discussed at length in later chapters, is that scientists tend to judge the adequacy of an experiment by its result. Thus, given a firm enough belief in, say, the non-existence of paranormal effects, a critic would tend to argue that any experiment which seems to demonstrate such an effect must be defective by that very fact. On the other hand, as we have just argued, it might be said that all critics were disqualified from producing results themselves unless they could first show their abilities by producing positive results; thus nearly all negative results would be ruled out of court. These kind of feedback loops complicate the assignation of success or failure to experiments.

An interesting problem, which may be special to parapsychology, concerns the nature of the 'positive result'. Parapsychologists have long agreed that the abilities of their subjects decline markedly after a time. It may be that the subjects get bored, or get tired, or lose their concentration. However, it can be argued that since the 'decline effect' is such a regular feature of experimental work in the area it, in itself, demonstrates indirectly that there is really something going on. If the entire phenomenon were an artifact, there would be nothing to decline! Such indirect effects have been called the 'fingerprints' of paranormal phenomena.[16] Another fingerprint is the 'sheep-goat' effect. The existence of this effect makes it possible to think of the long-term failure of some experimenters as a success for the programme as a whole! It is worth noting that during the debate among the parapsychologists Gaddum remarked that what he found convincing was that some people could see the phenomena while others could not.

Finally there is the problem of what level of statistical significance is to be taken to count as a positive result. As a rule social scientists are happy to consider a result positive if the likelihood of its arising purely by chance is less than five times in every hundred trials — the so called 'five percent level'. Such sciences are happy, effectively, if no more than one result in every twenty that are found in the published literature is wrong for predictable statistical reasons. Other sciences demand far higher levels of statistical significance if results are to be counted as fit for publication or public attestation. In most of physics a result is considered suspect if it is not so clear as to render statistical analysis otiose. There does not seem to be a rationale for the acceptability of different levels of statistical significance in different places.[17]

Level Seven: Let us accept that the philosopher-mice can squeak their way through all this and, invoking murine-rules as the need arises, arrive at a set of scientific experiments on topic 'r'. These have all been done competently by suitable investigators and unambiguously assigned with negative or positive results. Does it then follow naturally that 'r' has been replicated? There are three cases to consider: all results might be positive, all might be negative or there might be a mixture of positive and negative.

To start with the easiest, all negative, case the question would remain 'Have sufficient tests been done to warrant a conclusion?'. It may be that the tests have not been carried through with sufficient skill and determination. Though this problem has been reviewed at earlier stages, it has only been discussed in terms of the competence of individual experiments; each single experiment which has passed through to level seven has been adjudged competent by the mice, but since no single experiment can decide the matter, however well done it seems to be, the question of how many negative experiments ought to be done remains. After all, the whole problem is couched in terms of demonstrating the repeatability or non-repeatability of an observation over a series of tests. So how long must the series be?

On the other hand, if the results are all positive, it might look 'too good to be true'. Gaddum, as noted above, was impressed by variability of results, and if we accept the delicacy and fallibility of experimentation, it would be a surprise if everything worked out too perfectly. There are other circumstances, too, in which a string of positive results should not be expected; in the gravity wave case, to be discussed in Chapter Four, signals from galactic or extra-galactic sources were being monitored and such sources might be expected to fluctuate. In fact, Joseph Weber did argue that failures to confirm his findings might quite reasonably be explained by such fluctuations; the proper ratio of success to failure in detecting cosmic signals is far from clear and perpetual positive results might well be inappropriate.

Finally, suppose the outcome was a mixture of negative and positive. Can the result be determined? Actually, as will be seen, the value given to a single experiment, either a positive or negative, seems to depend upon the prior propensity of scientists to believe in the phenomenon in question. A frequently quoted example of apparent success for a single negative experiment is that done by R.W.Wood on 'N-Rays' (see, for example, Langmuir, 1953). French scientist Blondlot claimed that rays emanated from living matter and were readily visible under appropriate circumstances. Wood, invited to witness the rays, removed a crucial part of the apparatus in the dark — so the story goes — and looked on as Blondlot continued to 'see' them. This story, when it was reported, removed all credibility from the N-Ray work, at least outside France. Thus a single negative demonstration is said to have been worth all the positive results previously produced by Blondlot and his colleagues.

Compare this story of decisive disconfirmation with the oft-quoted myths of decisive confirmation of relativity through the single Michelson-Morley experiment and the single observation of the sun's bending of light rays by Eddington. Thus, the mice will have to adopt a criterion that is unknown to terrestrial creatures even at this final level of sorting.

At each level of the sieving process we have had to grant that the mice provided some solution to the problems and ambiguities which is unknown to us humans; in the terms of the story, we are not yet 'programmed' with it. And this should come as no surprise. The humans in the story are working in 'open systems'. What is more, in doing science they are *developing* the rules for 'carrying on in the same way' rather than applying them. In other words, they are developing the conceptual system rather than using it. In Chapter One the problems of acting in an orderly fashion — the problems of seeing similarity and difference — were explored. The ability to cope in open systems is precisely what we were unable to explain. The surprise, then, should not be that Terraputer (terrestrial science) runs sluggishly in certain areas but that it runs at all. It is not disorder that is mysterious but order, that order which is so manifest in nearly every human activity.

In establishing and maintaining order in new areas of scientific endeavour humans somehow cope with this open-system. One aspect of the establishment of this order is the agreement about what phenomena are to be taken as replicable. In effect, humans successfully negotiate their way through the sorting levels and agree to demarcation criteria at each level without access to the murine rules. The murine solutions are opaque to us — we live successfully in open systems even though we cannot analyse or formulate their rules of action.

The project of (some) philosophers of science is to explicate the rules. Those philosophers with prescriptive ambitions want to explicate them in order that the progress of science can be more effective. Their project is to write the programme of the philosopher-mice in the story.[18] One cannot but applaud this effort to uncover the rules of proper scientific behaviour. If the search were successful — if the murine rules could be discovered and explained — then it would be possible to run science on Earth just as the mice in the story wanted; it would be possible to provide a programme, like the programme of a digital computer, for the correct pursuit of science. Each actor/component would then live out a predictable role in the sense that the behaviour of each component in a digital computer is predictable.

On the other hand, in Chapter One we saw that the goal of artificial intelligence research is to build a computer which acts like a human in an open system. Clearly the project of rationalist philsophers and the project of the 'artificial intelligencia' (as artificial intelligence researchers are known) converge. One group wants to discover the formal programme of human action while the second group wants to write a programme that will mimic human action. The programme that both seek is essentially the same one, though they approach it from opposite sides. Picturing human science from the murine perspective — so that humans are seen as components in a great computer — makes the convergence of the two projects clear.[19] The identity of these two problems suggests that workers in both fields have much to learn from each others' successes and failures.

The next chapters examine scientists working in practice. The first of these chapters establishes and explains my claim that experiments are delicate and fallible. The following two chapters look at attempts and methods of developing rules for what is to count as 'going on in the same way' in new areas of science.

Notes

1. Popper is known as an arch 'anti-inductivist'. The basis of his influential philosophy is the distinction between the 'impossible' process of corroboration of a theory by repeated observation of its consequences and the falsification of the theory which can be done on the basis of only one instance. For example, the theory 'all swans are white' can never be proved however many white swans are seen, but it can be disproved if only a single black swan is seen.

Popper's philosophy, though it maintains an irresistible appeal, has been widely criticized on a number of counts. For the present, we need only note that the process of falsification would not allow us to gain any knowledge of the world unless we start off with a small number of reasonable theories. If we start with the indefinite number of possible theories, then falsifying a few of them will not get us sensibly nearer to the truth. Thus Popper does not avoid the problem of induction. Inductive processes must be what give us the small number of reasonable theories in the first place.

2. Rosenthal had done experiments in which different groups of experimenters were told to expect different outcomes to the identical experiments that they were doing. Their respective results followed the direction of their expectations.

3. It ought to be noted that the most readily replicable result discussed in these quotations, the result of Soal, has now been thoroughly discredited. It seems that Soal cheated.

4. A recent article by Franklin and Howson (1984) served as the starting point for this part of the chapter. See also my reply (Collins, 1984b).

5. This follows clearly from the idea of expectancy effects.

6. Actually, who counts as a 'proper' experimenter varies from field to field. In 'low status' areas, such as plant perception (discussed in Chapter Five) graduate students' work and even high school students' work may be cited as providing legitimate evidence one way or another, even though it would never be taken seriously in the prestige areas of science. Similarly, television personnel often take it upon themselves to offer substantive experimental or theoretical commentary on areas of low status science.

7. Lakatos's work is fascinating. He shows that the asymmetry between corroboration and falsification proposed by Popper (see note 1) is far less clear cut than it appears. Given an apparent falsification of a hypothesis, a determined upholder can always put forward sub-hypotheses to rescue the phenomenon. For example, take once more the theory 'all swans are white'. If a black swan is seen this does not necessarily falsify the theory because the upholder might argue that the swan had been covered in black paint, or wasn't really a swan at all.

In his most celebrated study (1976) Lakatos reconstructs the history of Euler's theorem — a theorem about the relationship between the number of sides and the number of vertices of a polyhedron. He sets out the history as a series of arguments between those who wanted to uphold Euler's claimed relationship and those who wished to disprove it. The tactics for upholding an hypothesis in the face of apparent counter-evidence are brought out, and the various possible moves are labelled. For example, very oddly shaped polyhedra, which do not obey Euler's relationship, might be disbarred on the grounds that they were 'monsters' by those who wanted to rescue the hypothesis. Lakatos calls this 'monsterbarring'.

The aspect of his work that we refer to here is the less successful positive side (see, for example, Lakatos 1979). Here he attempts to distinguish between reasonable and unreasonable rescue hypotheses. He divides up programmes of research into a 'hard core' and a softer periphery, and he suggests that scientists may legitimately sacrifice elements of the periphery, but must not sacrifice the hard core of the programme or they sacrifice the programme itself. While this is intuitively appealing — that is why I make a similar move in the analytic theory of replication — it does not work because one does not find uniform perceptions of what counts as core and periphery among scientists. It is only with the benefit of hindsight that one can see what was important and what was not.

8. The exact position of the dividing line between science and pseudo-science is a nice question. It is especially problematic in Figure 1a where the vertical line might well

be drawn somewhat to the right of where it is shown. Really 'cranky' science can produce almost zero confirming power while suspect orthodox science produces little.

In the case of the controversy over the detection of gravitational radiation, which I will deal with in later chapters, an Israeli scientist, Dror Sadeh, claimed to have detected the influence of gravity waves in the vibrations of the Earth. (Other scientists had used artificial antennae.) Sadeh's results, however, had almost no confirming power even though he was an accredited physicist using sophisticated experimental techniques. Other scientists felt his work was suspect, believing that he had made incorrect claims before, and believing the Earth to be an insufficiently noise-free detector to see the elusive waves. The crucial point, as will be argued in later chapters, is that scientists cannot reach consensus over the power of experiments until the nature of the proper outcome of a proper experiment is agreed.

9. Rosenthal (1966) is particularly interesting on the relationship of confirmatory power and social proximity.

10. This is a modified version of an idea first put forward in Collins (1976). Level Two is a new addition. In the earlier version the levels were referred to as stages, but I have renamed them 'levels' to save confusion with the stages in my 'Stages in the Empirical Programme of Relativism' (1981b). Stephen Braude (1980) adapted the original scheme (with appropriate acknowledgement) to form the basis of a chapter in his interesting work on the philosophy of parapsychology.

11. One staunch critic of parapsychology, Professor C.E.M.Hansel, remarked to me that he would never trust any finding that he could not replicate *himself*. This is carrying the principle of socio-cognitive differentiation to a novel extreme. It would be disastrous were it to be applied to science as a whole. Imagine every physicist demanding that he or she be allowed to replicate every finding.

12. In parapsychology it is probably true that most scientists come to believe in the phenomena as a result of some personal experience or as a result of others' personal experiences, the sum of which are termed 'spontaneous phenomena'. Nevertheless, the elusive validation experiment is still pursued with vigour.

13. Actually, parapsychologists have an answer to this criticism. They argue that the aggregate positive statistics on their experiments are so high that, even if all the scientists in the world had spent all their time doing negative experiments since pre-historic times, they still could not have enough negative experiments in their file drawers to counterbalance reported success (eg, see Tart, 1973). This does not affect the point of principle being discussed here. In any case parapsychologists now take enormous precautions in the way they record every experimental run and report even negative results so as to render the criticism inapplicable. My own view is that these precautions are exaggerated; public education in the matter of the fallibility of experiment is what is needed.

14. An undesirable complication enters here since Rosenthal is involved in trying to discount expectancy effects on experiments designed to look for expectancy effects themselves. Ignore this complication.

15. For a variety of reasons I am sure Rosenthal's expectancy effect work is correct. However, I believe that his aggregate statistical arguments for it are unconvincing.

16. See, for example, Beloff (1982). In parapsychology there is another interesting problem. In statistical experiments — where the subject is to guess a sequence of unseen symbol cards — it is easy to predict the number of correct guesses that should arise if purely random factors are responsible. For example, if there are five different symbols, the subject should tend toward one fifth correct guesses in the long run. The usual aim, and usual outcome of successful experiments, is that the subject guesses significantly more than one fifth.

Sometimes, however, subjects get significantly less than one fifth correct guesses, an outcome equally unlikely according to the laws of chance. How is such an outcome to be counted? Is it a negative result, or is it a *successful* case of paranormally mediated *guess missing*? The answer is important because the combination of a successful hitting experiment and a successful missing experiment add up either to two successful experiments or to a precisely null result! The problem is usually avoided by stating in advance what the aim of the experiment is to be, but given the delicacy of experiments, this is not an entirely satisfactory state of affairs.

17. For a fascinating discussion of the rationale for the uses of statistics in psychology see Henkel and Morrison (1970). For a sophisticated account of the development of certain statistical tests, revealing the interests of those who developed them and the way this affected decisions regarding the formulae that were finally accepted, see MacKenzie (1981).

18. Their aims contrast with other philosophers who have accepted that the attempt to explicate the rules of human action merely creates puzzlement.

19. To press the fable to its limits we could think of the creation of the 'artificial intelligentsia' as one of several attempts by the mice to get Terraputer to find out what was wrong with itself and why it ran so sluggishly and uncertainly. We can imagine that the first attempt was to develop a psychology section within the programme to try to discover the faults in the individual components. The second attempt, it may be imagined, was to develop a philosophy section to try to solve the problem at the analytic level. The third attempt was an effort at an experimentally based solution where Earthlings try to build models of their own Terraputer — artificial intelligence. True understanding, I suggest, awaits the development of the sociological component.

Chapter Three

Replicating the TEA-Laser:
Maintaining Scientific Knowledge[1]

It is the 'easy' cases of replication — cases where the murine rules present few problems and where orderly action is most readily achieved — which are the most mysterious. So long as it is thought that easy science is really an easy matter, it will be hard to see it as the social accomplishment that it is. For this reason I start the empirical work of this book with an analysis of the case of laser building, a piece of straightforward 'normal' science where no one doubted that the phenomenon could be replicated.

The TEA-laser

A laser produces a beam of powerful 'coherent' radiation, often visible light, that can be focused very finely and can therefore damage the small spot upon which it impinges. The radiation is generated by putting energy into the molecules of the lasing substance — it might be a piece of ruby, or a gas — and then releasing all this energy in a synchronized way. The TEA-laser uses a gas as the lasing medium and produces infra-red radiation rather than visible light. If properly focused this radiation can vaporize concrete or burn the silver from a mirror. However, the crucial part of this story turns not upon what happens after the gas molecules give up their energy, but on how these molecules are energized in the first place.

The TEA-laser uses Carbon-Dioxide (CO_2) as the lasing medium. This is mixed with quantities of helium and nitrogen. The gas is held in a glass or perspex tube and is energized by passing an electrical discharge through it. When a gas is energized in this way it glows. A neon display tube, such as is used in shop signs, is a tube of electrically energized neon gas. The colour of the glow depends on the nature of the gas. Red is the characteristic of neon whereas the TEA-laser gases produce a pleasant pinkish/whiteish/blueish glow.

A standard gas laser is like a neon display tube in that the enclosed gas is energized by passing a high-voltage current through it between electrodes placed at either ends of the tube. A laser is made by using the appropriate gases and voltages and making suitable arrangements of optical windows and mirrors at the ends of the tube. With such an arrangement, however, a uniform glow discharge can only be obtained in lasers (and display tubes) if the gas is at a very low pressure, a very small fraction of atmospheric pressure. As the gas pressure inside such a tube is raised it is harder and harder to 'force' the electricity through the gas. Higher and higher voltages are required, and at 'high' pressures the current will only pass along a narrow path and only with a sudden jump. This is what we see as a 'spark', or an 'arc' breakdown, of which lightning is an example on the largest scale. But a glow discharge is needed for a laser.

The power of a gas laser is proportional to the amount of gas that can be energized, and the amount of gas in a container of given size is proportional to its pressure. Thus the power of gas lasers was initially limited by the low pressure of the lasing medium made necessary by the need for a glow discharge. It is this barrier that the TEA-laser broke through. The TEA-laser uses gas at atmospheric pressure. It generates a glow discharge in this 'high pressure' gas by using very high voltages, by placing the electrodes at either side of the tube rather than at either end, so that the path between them is shorter and so that their area can be increased, and by discharging in pulses rather than continuously. That is why it is called the Transversely Excited Atmospheric pressure CO_2 laser. The difficulties in building such a device lie in the electrode structure, the electronics which shape the pulses of electricity, and the high voltages. These will be the main features in the discussion which follows.

Appearance and construction

The gas tube on the TEA-laser models that I saw ranged in size from about three aerosol cans placed end to end up to the size of a golf bag. The small, early devices used round glass tubes, whereas 'Jumbo', one of the more powerful lasers to be discussed later, used a perspex box of square cross section. Since the gas was at, or fractionally above, atmospheric pressure no special care needed to be taken in preparing the gas vessel against leaks, implosion, and so forth. Indeed, the first article in *New Scientist* to reveal the existence of the laser drew attention to the relative simplicity of the gas vessel by giving it the headline 'plywood-laser'.

The electrodes (the positive 'anode' and the negative 'cathode') are easily visible, running along either side within the tube. The design of these changed considerably as the state of the art developed between 1969 and 1979. The early models were known as 'pin-bar' lasers because one electrode took the form of a plain bar, and one a series of pins. Later devices, such as Jumbo, used artfully curved plates of metal at one side, and a flat or finned plate with a row of 'trigger wires' just above it on the other.

Electrical Circuit Diagram for the TEA CO$_2$ Laser

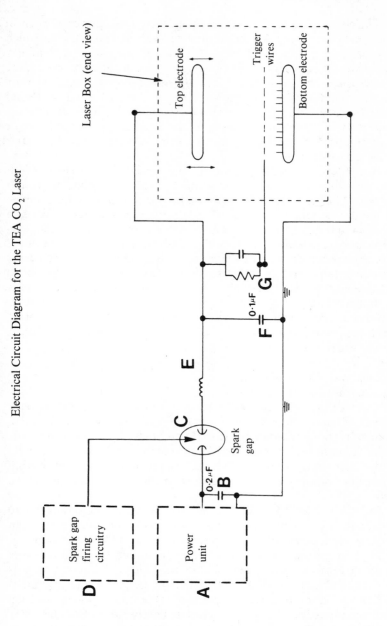

The electrical system of such a device includes a 'power unit' (A in the figure above) capable of building up to around 60,000 volts which is usually an 'off the shelf' item. The power unit charges a 'primary-capacitor' or capacitors (B in figure above). The latter are required to dump their charge across the two electrodes once they are fully charged, and for this purpose a special 'spark gap' switch is needed (C). This is itself a small gas filled tube that can be made to conduct electricity by creating a spark within it. An ordinary automobile spark plug can be used for this purpose but some additional circuitry (D) is required to fire it.

The pulse of electricity dumped by the capacitors has to be the right 'shape'. For example it has to rise quickly and steadily and not be too 'jerky'. The pulse is shaped by being passed through an 'inductance' (E) and a 'secondary capacitor' (F). In more sophisticated lasers a portion of the charge goes to the trigger wires through the 'resistance-capacitor (rc) circuit' (G). The figure above represents a simplified circuit diagram for one of the more sophisticated lasers.[2] The cross section of the top electrode in such designs is shaped in what was called a 'Rogowski profile'. Designs such as this, which start their discharge with a small trigger pulse between the trigger wires and the bottom electrode, are known as 'double-discharge' lasers. The triggering pulse is designed to generate a small area of 'pre-ionization' of the gas within the tube.

Replication of the early lasers:
the transmission of knowledge

Early in 1970, when no one else had yet achieved successful gas laser operation at pressures above about half an atmosphere, a Canadian defence research laboratory, which I will call 'Origin', announced the TEA-laser. In fact, the device had first been operated early in 1968, and a more sophisticated version had been built by the autumn of that year; but both generations of laser were classified as secret for two years.

In 1971 and 1972 I decided to talk to scientists who were trying to build copies of the device in Britain, and to find out what they did in order to replicate the original experimental finding.[3] In the summer of 1971, I located seven British laboratories which had built or were building TEA-lasers; I visited six of them.[4] This was eighteen months after the first news of the device came from Origin. In the autumn of 1972 I visited the five North American laboratories which had been involved in the transfer of laser building knowledge to British sites.

The seven British laboratories comprised two government-run laboratories and five university physics or applied physics departments of physics or applied physics. The five North American laboratories comprised two government-run establishments (both Canadian), one American university department and two separate research laboratories belonging to the same American firm.

I found that the transmission of the ability to build a TEA-laser was not a straightforward matter. The flow of knowledge between the laboratories was constrained in a number of ways. There were some constraints, which are of interest but of only small significance for this study, which emerged out of what seemed to be competition among laboratories. Thus many communication links that could have proved useful to the less advanced centres were never realized, even though their potential was understood (Collins, 1974). A second constraint caused by competition affected the communication links that *had* been actualized. In some cases the knowledgeable institution would not be completely open with members of the learning institution. Thus one scientist reported of a visit to another laboratory:

> They showed me roughly what it looked like but they wouldn't show me anything as to how they managed to damage mirrors. I had not a rebuff, but they were very cautious.

A more subtle tactic used was that of answering questions, but not actually volunteering information. This maintains the appearance of openness while many important items of information are withheld; their significance will not occur to the questioner. One scientist put it:

> If someone comes here to look at the laser the normal approach is to answer their questions, but... although it's in our interests to answer their questions in an information exchange, we don't give our liberty.

Another remarked succinctly:

> Let's say I've always told the truth, nothing but the truth, but not the whole truth.

The more significant constraints, as far as this book is concerned, operated where there was no conscious attempt to conceal information. The first point is that no scientist succeeded in building a laser by using only information found in published or other written sources. Thus every scientist who managed to copy the laser obtained a crucial component of the requisite knowledge from personal contact and discussion.

A second point is that no scientist succeeded in building a TEA-laser where their informant was a 'middle man' who had not built a device himself. The third point is that even where the informant had built a successful device, and where information flowed freely as far as could be seen, the learner would be unlikely to succeed without some extended period of contact with the informant and, in some cases, would not succeed at all. The extended contact might come through exchange visits of laboratory personnel, or regular cooperation, or a series of visits and telephone calls. Typically, a laboratory visit might be followed by an attempt to build a laser which would not work, so another visit would follow, and if success was still elusive, a telephone call, or perhaps several, would follow. In at least one case, even this type of sequence resulted in failure, and the unsuccessful laboratory eventually abandoned its attempts to build a device.

In sum, the flow of knowledge was such that, first, it travelled only where there was personal contact with an accomplished practitioner; second, its passage was invisible so that scientists did not know whether they had the relevant expertise to build a laser until they tried it; and, third, it was so capricious that similar relationships between teacher and learner might or might not result in the transfer of knowledge. These characteristics of the flow of knowledge make sense if a crucial component in laser building ability is 'tacit knowledge'.

Tacit knowledge
Tacit knowledge is the name given by Michael Polanyi (1958, 1967) to our ability to perform skills without being able to articulate how we do them. The standard example is the skill involved in riding a bicycle. No amount of reading and study in the physics and dynamics of the bicycle will enable a novice to get on and ride immediately. On the other hand, the skilled rider is usually quite unable to describe the dynamics of balance involved. Does one turn the handlebars to the right when one feels oneself falling to the right? Or is it that one shifts one's weight when one senses approaching disequilibrium? The rider simply does not know. All the rider does is 'ride a bike'. The experience of riding rarely involves the anticipation of an imminent fall which must be avoided by a deliberate act of balancing (except perhaps at very low speeds). Even the early learning process does not seem to be aided by attempts to articulate what is required to stay balanced. It is a matter of trying over and over again until the skill, whatever it is, has been mastered.

Tacit knowledge usually finds its application in practical settings such as bike riding or other 'skilled' occupations. However, it is equally applicable to mental activity. Thus, to return to an earlier example, the member of a social group who has the ability to continue the sequence '2,4,6,8' with '10,12,14,16' as a matter of course, without even thinking about it, also possesses something that the stranger to our culture and the newborn do not. This is sometimes referred to as 'social skill' but we can call it tacit knowledge without doing too much violence to the term. It forms the foundation upon which formal learning rests. If I am taught some new algebraic manipulation in school, and the teacher tells me to do it the same way next time, I can say that it is my tacit knowledge which tells me what counts as the next instance of the same problem as well as what is meant by proceeding in the same way. (Remember the fundamental ambiguity of such an instruction as discussed in chapter one.)[5]

Two models of learning

This discussion suggests two models of learning. One model rests upon a notion of knowledge as a set of formal instructions, or pieces of 'information', about what to do in a variety of circumstances. This model views knowledge as the sort of information that enables a computer to carry out its programmer's intentions — I will call it the *algorithmical model*. The other model looks upon knowledge as being like, or at least based on, a set of social skills. It is what the child or the stranger must know before they understand what it means to go on in the 'same way', whether the same way is what is required at a cocktail party or required of an audio-typist or required of a member of the communities of physicists, mathematicians, parapsychologists or laser builders. This I call the *enculturational model*.

If a crucial component of laser building ability is tacit knowledge, then it should come as no surprise that written information turned out to be an inadequate source. Likewise, one would not expect that a 'middle man', who had not mastered the skill himself, would be able to pass it on. Further, since laser building skill, just like bicycle riding skill, is invisible in its passage and in its possession scientists who thought they knew how to build the laser discovered that they did not know how — this is no more surprising than if an expert in dynamics, having never ridden before, were to fall off a bicycle. Finally, it ought not to be surprising that the passage of the skill is not *completely* determined by the extent of the personal contacts between scientists; after all, as with other skills, the most lengthy training does not guarantee mastery. All these are the predictable consequences of the enculturational model of learning and communication; they do not follow from the algorithmical model.

That there was a tacit component of TEA-laser builders' knowledge was apparent to some builders. Thus the invention and state of knowledge concerning the 'Double Discharge' laser in 1972 was described to me by its inventor as follows:

> First of all we had rows of fins instead of pins, but this didn't work too well. We thought this might be because the field uniformity was too great so we put a row of trigger wires near the fins to disturb the field uniformity. Then we started finding out things. It did improve the discharge but there were delays involved. It definitely worked differently to the rationale we had when we first made it ...
> Even today there is no clear idea about how to get this thing working properly. We are even now discovering things about how to control the performance of these devices, which are unknown ...

> I have four theories (for how they work) which contradict each other
> ...The crucial part (in getting a device to operate) is in the mechanical
> arrangements, and how you get the things all integrated together. In the
> electrical characteristics of the mechanical structures ...This is all the
> black art that goes into building radar transmitters.

Again, the misleading quality of some of the formal information available in 1972 can be seen in different laboratories' beliefs and actions regarding the proper shape of the electrodes: one source laboratory provided information in the form of a set of equations for the so-called 'Rogowski profiles', along with the impression that machining tolerances must be small. The difficulties involved in making the electrodes were found by another laboratory to be insuperable. In the meantime, another British, laboratory had produced the shapes roughly from templates and a filing operation, while an American laboratory had simply used lengths of aluminium banister rail, both with complete success. The capriciousness of the knowledge flow is obvious, since even those who had succeeded in building a laser and making it work did not fully understand it!

Laser building in 1974 and 1979

What follows is a far more detailed examination of one scientist's attempts to replicate another laser. A physicist and expert in non-linear optics, Dr Bob Harrison (then at the University of Bath) set about building a TEA-laser of the double discharge design in early 1974. Harrison had previous experience of working with a laser of this design, and he had excellent contacts with a major laboratory where similar devices were in regular use. He visited this laboratory regularly. I was able to persuade him to keep a diary of his work and I regularly visited his laboratory and helped out with work on the laser.

Harrison eventually moved to another university, taking his working laser (nicknamed Jumbo) with him. In 1978−9 he built a near identical copy of Jumbo and I was able to be present, and help out in the crucial last session of development, from the first trial nearly up to the moment that the device finally worked. A description of this passage of work forms the latter part of this chapter.

Building Jumbo

In spite of Harrison's experience and excellent contacts it took him six months from the assembly of the parts to the final ironing out of the faults to make Jumbo work. There were some uncontrollable outside delays, such as those involved in the return and repair of faulty manufactured parts and repairs to a leaking laboratory roof. However, a large part of the time was spent on 'debugging' the device, or as I would prefer to say, developing the relevant tacit skills. A full account of the building of Jumbo has been published elsewhere (Collins and Harrison, 1975) and I will report only that part of the work which bears on the major thesis of this book.

Jumbo had parts in common with the lasers that Harrison had worked with before. Indeed, the laser cavity and one of the electrodes was supplied by Harrison's contact laboratory. Jumbo was intended to differ in design from these models only so as to make the layout of the electrical components slightly more tidy, since Harrison wished the whole unit to be easily portable. Thus, the high voltage components were arranged on a moveable trolley beneath the laser cavity itself. Harrison (henceforward 'H') described the general principle of the lay-out as follows:

> ... with high voltage stuff, keep everything well apart ... I knew working orders — breakdown in air is 30KV per centimetre at atmospheric pressure — that plus a kind of intuitive feel. But that's a hard thing to work out — is it breakdown between two flat surfaces or two points? — so actual distances are much bigger than this rule of thumb would imply ... I just made sure that I had a 500% or 1,000% safety factor ... Just keep things well spaced out — who needs trouble?

Arc breakdowns among the components

The first set of problems to be encountered was to do with arc breakdowns (enormous sparks at around 60,000 volts) among the various electrical components in the trolley. In particular, some 'arcing' occurred between earth points and other earth points which should have been at the same — zero — voltage!

> For example, I would have from the capacitor earth connection, an earth lead, and that lead would perhaps touch the housing of the capacitor which of course is meant to be at earth, and there would be a breakdown between the earth lead and the capacitor housing where it was touching, as far as one could see.

H also found arc breakdowns between components separated by much more than one centimetre for every 30,000 volts — the 'rule of thumb' distance. H knew that this could be explained if the components had sharp points, giving rise to 'dark field emissions' which may eventually break down suddenly when the voltage gets high enough. The dark field emissions prepare a path for the arc breakdown by pre-ionizing the air.

> But you 'wave hands' at this point because it's a subject that's been going on a hundred years and is still pretty difficult to understand, except under controlled physical conditions where, say, you have two flat surfaces and perhaps a point. But where you have a few curves and edges and you're not really looking at it, anything can happen ... but remember when we had that really enormous breakdown, that was from a High Tension lead to

earth, and that just didn't make sense rationally because there was a bloody foot separating them, or something ridiculous, and it looked as though it was arcing to wood!

H solved these problems in the most pragmatic way: wherever there was a breakdown between earth and earth he insulated with polythene sheet and wherever there was a breakdown between a high tension component and earth he covered any points and edges with cut up polythene bottles. I helped in this process of firing the capacitors, spotting the devastatingly loud and bright breakdowns, insulating, firing again, insulating again, and so on. Eventually, H had to drape most of the components in polythene sheet.

H discussed the earth-to-earth breakdowns with a colleague, and they agreed that it was probably a phenomenon associated with 'transient currents'; these could arise from differences in potential between common conductors where the rise time of a current pulse is very fast.

> It's intuitive evidence — 'beware my lad when you go to short, high voltage stuff — transients and odd things creep in'. It's word of mouth. It's a technology that's evolved, people just give you guidelines and say, 'well, this is to be expected', without really full qualifications ... [the transient currents hypothesis] proved to be sensible but clearly only sensible because there was no other plausible reason.

Probe coil

When at last this programme of trial and error insulation had succeeded, H encountered problems within the laser cavity proper. He could not achieve the desired glow discharge, only sparks and arcs. After trials with different electrode separations he concluded that the pulse profile (see above) must be wrong. He therefore decided to monitor the shape of the discharge pulse, using a technique he had picked up earlier. This involved probing in the area of one of the high tension leads with a small inductive coil monitored by an oscilloscope. H soon found that his coil picked up so much radio frequency noise from the laser that the pulse shape information was completely masked. He spent some time moving the probe coil about, shielding the wires, using a Faraday cage on the oscilloscope, and so on; eventually he had to give up, as none of these precautions could sufficiently reduce the noise. At this point H remarked that he was:

> ... fairly desperate — I thought 'where the hell can I go from now?' I rang up 'D' [H's main contact] and he said 'it's a joke'. There are enormous problems trying to do it [the probe technique] ... Although — the bugger — I'd rung him up before and he'd told me how to do it and hadn't told me that it was a bloody virtually impossible thing to do. I'm sure of this — so I had a go and couldn't believe the things I was getting.

H then decided to visit his contact laboratory.

Leads and tubes

> I was then in a situation of — well, what shall I do with this damn laser? It's arcing and I can't [monitor the discharge pulse] so I thought well, let's take a trip to [the contact laboratory] and just make sure that simple characteristics like length of leads, glass tubes [these are part of the lower laser electrode] look right. If they're hopelessly wrong let's correct them.

H knew that the leads from the capacitors to the electrodes had to be short, and the glass tubes flat, but had not given any quantitative consideration to these matters. In designing and building his laser he remarked that he:

> ... had to support God knows how many pounds of capacitors — I knew that they had to be close, just how close I hadn't really [bothered to think about too seriously], so I put them as close as they could comfortably go in an upright position that was convenient for housing them: so, it was convenience — I know they've got to be close, at the same time we don't want to frig around with too much framework...

These leads were about eight inches long in the laser as first built, which, as H remarks, is 'short by any standards'. As regards the glass tubes, H knew they had to be flat from his earlier days. He says:

> In fact I had a communication way back in those days from a chap in Livermore, who was over here working, not in TEA-lasers but knew a few people who were, and he sent me some information. At that time ... we didn't know how flat, we knew flat, and he wrote back to me giving me all these values, and the implication was they had to be incredibly flat. So this was always in the back of your mind — get things as flat as you can get them. That was really the criterion all along.

When H went back to the source laboratory, he noticed things he had not 'seen' before. He found that their capacitor leads were considerably shorter than his, and there was 'no limit to how short they should be, just as short as possible'; this had involved the scientists in inverting their capacitors to reduce the lead lengths. H had not noticed this before. As he said, there was no reason why he should be building one exactly like that of the source laboratory down to the detailed positioning of the electronic components.

Upon H's return he took the bottom electrode apart to check on the glass tubes, which he had seen were very flat indeed in the functioning model. He found that his tubes did not fit the bottom electrode properly: they were slightly too large, and thus were not held properly flat in their wells. After some trouble, he managed to fit flatter tubes in the bottom electrode, and to have the capacitors mounted upside down so that the leads were still shorter.

When the laser was reassembled, H tested it again and found that it was *still arcing* between the electrodes. All these modifications had not cured the basic problem.

Anode marking

H then telephoned his contact laboratory for advice on another troublesome problem to do with the spark gap. He chanced to remark that the arc discharges between the electrodes were marking the anode; he knew that other TEA laser systems left marks on the electrodes, and was not at all surprised to find the effect on his anode, so he added this comment 'by the way'. However, it prompted the experimental officer to suggest that H check the polarity of his power source; the officer had seen marks on the anode on an occasion when the electrodes were accidentally connected up the wrong way round.

H didn't completely dismiss this possibility, though he thought it rather unlikely; nevertheless, a quick check with a meter showed that in fact his power unit was delivering +60,000 volts, instead of -60,000 volts. Upon rearranging the connections to the electrodes and sorting out a few minor arcing problems, H found that he was at last able to obtain the desired glow discharge in the laser cavity.

H remarked at the time that if it had not been for a lucky telephone conversation he would certainly have continued to spend time and effort checking blind alleys, perhaps until some other fortuitous event caused him to notice the elementary polarity-reversal error.

This account confirms the findings of the network study regarding the nature of communication of TEA-laser building ability. It is difficult to explain H's problems by reference to any shortfall in his information sources. His informant laboratory was a place where he had worked; what is more, he had worked there at the early stages of the very same design of laser that he was now building. He maintained a continual consultancy relationship with the laboratory, visiting them regularly. And, far from being competitive or secretive, his source had loaned him some £1,500 worth of equipment to help him build the laser. Nevertheless, deliberate transfer of the requisite knowledge proved extremely difficult.

It is clear that there were long periods when, though he did not have laser building ability, *H did not know* that he did not have it except by reference to the fact that the laser did not work. During these times he expected Jumbo to function but it did not. In the end, one crucial piece of knowledge seemed to be only accidentally transferred. Sometimes these failures of communication seem to be simply a matter of poor understanding of the laser parameters — length of leads, flatness of glass tubes — and sometimes it seems that H was just making mistakes — polarity reversal. There is little doubt that H felt rather silly that his laser had failed because of such an elementary mistake and yet, as he said at the time, it is probable that most of the other things would have had to be set right anyway, so little time was lost on the polarity reversal alone. Nevertheless, as we shall see, being 'wise after the event' is an almost inescapable feeling in scientific work.

Building the Heriot-Watt laser

Bob Harrison started to build his second TEA laser at Heriot-Watt University, in Edinburgh, in the middle of 1978; he was ready to test fire it for the first time at 12.10 pm on 15 March 1979. I was able to be present at Heriot-Watt on 15 and 16 March; this turned out to be the whole period between first testing the laser and the moment of successful operation with the exception of the final two hours work. These were done on the morning of 20 March. Bob Harrison gave me an account of these two hours by telephone on the same Monday afternoon.

The material presented here comes then from an initial telephone discussion in January 1979, two days spent in the laser laboratory at Heriot-Watt recording events by tape recorder and notebook, tape recorded discussions on the evening of 16 March and notes made from a telephone call on 20 March. During the 15 and 16 March I was able to participate, and make occasional useful suggestions.

The account here is very unusual from the point of view of the knowledge transfer perspective. H had already built Jumbo, so the relationship between teacher and learner was one of identity; they were both Bob Harrison. There was certainly then, no shortfall in H's sources of information! The other unusual feature is that H had a working laser — Jumbo — alongside the new one he was building. Jumbo had been performing reliably for a number of years in H's new laboratory. Thus, immediate comparisons were readily made between the old and the new devices and parts could be easily interchanged between the two.

Finally, and this again makes the setting particularly pertinent to the thesis of this book, H intended to make the new laser as nearly as possible the same as Jumbo. The only difference that H wanted to build into the new laser was a higher repetition rate for its pulses of power. Jumbo could produce a pulse every six seconds but the new laser was to produce a pulse every two seconds. This difference was to be accomplished by changes in 'off the shelf' units.

The reader should now turn back to the explanation of the laser, and the simplified circuit diagram given in the figure. The major components in Jumbo and the new laser were all discrete and visible large objects; their size and separation was made necessary by the very high voltages being used — up to 60,000 volts. All except the secondary capacitor were installed in a metal box about six feet long and four feet square upon which the laser cavity and gas container rested. The secondary capacitor was housed alongside the laser box so that the leads from it to the laser main electrodes could be as short as possible.

The laser itself, marked in the dotted square to the right of the circuit diagram, consisted of a perspex box about five feet long and one foot square section with gas inlet and outlet. The top electrode was a bar of aluminium about four feet long, six inches wide and half an inch deep milled to a 'Rogowski profile'. The bottom electrode was the same length but about an inch narrower, and had eight deep grooves milled in its upper surface running the length of the electrode. In these grooves sat eight glass tubes containing 'trigger wires' (marked in the circuit diagram as dashes above the bottom electrode). The top electrode could be adjusted until it was exactly parallel with the bottom one. So far the new laser was the same as Jumbo except that in the former there were two secondary capacitors side by side instead of one. With improvements in capacitor technology the same capacitance can be fitted into one unit.

The laser works as follows: the power unit charges the primary capacitor up to a potential which is pre-set on a dial — for example, 45,000 volts. On command (this is what pressing the button amounts to), the spark gap is fired and the primary capacitor charges the secondary capacitor through the inductance. This in turn is supposed to discharge across the laser electrodes.

As explained above, a uniform discharge rather than an arc is required. To accomplish this, part of the charge in the secondary capacitor is fed to the trigger wires through the resistance-capacitance (rc) circuit; this should pull electrons from the bottom electrode, which will 'pre-ionize' the gas in the cavity, facilitating uniform main discharge. In these circumstances, pre-ionization is visible as a faint pinkish glow just above the bottom electrode, and the main discharge fills the space between the electrodes with pinkish light. The laser then makes a sound like a loud 'ping'. An arc discharge makes a blinding flash at one point in the box and a ringing 'crack'.

H had deliberately built the new laser to resemble Jumbo as closely as possible in all essentials because he wanted to waste as little time as possible on irrelevant details. However, he had bought the most suitable 'off the shelf' components currently available. Thus, both capacitors in the new model were much smaller in size for the same capacitance, the spark gap was very much smaller; and the power unit was different as was the spark gap power supply with its transformer. The perspex box of the laser was identical to Jumbo's and both had been supplied by H's contact laboratory with whom he retained good relations.

Similarities and differences

When I arrived at Heriot-Watt at 10.30 am on 15 March I discussed these similarities and differences with Bob Harrison. The point was to determine *what counted as a difference* for H and what might subsequently be taken to account for failure of the laser. We noted the fact that the gap between the electrodes was the same in both lasers to the fraction of a millimetre. We remarked that in the new laser the tungsten wires emerging from the glass rods had snapped off, leaving only very small stubs which had been connected to their joint cable with electrically conducting glue; in Jumbo the wires were screwed into a brass connecting block. H remarked that he was not sure if the new method would work but that it should do so. Another change was that the glass rods in the new laser were held in place by elastic bands rather than a perspex clip. This seemed unlikely to affect performance. The connections to the electrodes were different, too, with only one connection to each electrode on the new laser, but four to each Jumbo.

H commented:

> ...so that could be slightly different again because we still really don't know whether it's important that you make only one connection or whether you try to distribute the charge across the whole electrode.

H remarked that the new capacitors, though smaller, were likely to work better. However, he also remarked that he was 'very suspicious' of the very small new spark gap and he did not think that it was going to work. He pointed out that the complex spark gap in Jumbo had been replaced with a single box, plus a tiny transformer to convert 350 volts to 35,000.

Another difference pointed out by H was that the new laser's bottom electrode had been made in two parts and bolted together, but he did not think that this would be important.

> That bottom electrode may be a bit worrying because its made in two sections — look, you can see the joint — and we've just got to see whether that's going to cause arcing. I suspect it won't. I think that will probably be all right.

I asked about the flatness of the glass tubes which at one time had seemed to be a crucial parameter in Jumbo, but H seemed unconcerned about this. Finally he commented:

> Everything else is identical ... and [jokingly] I guarantee that it won't work first time.

After some discussion we checked on the gas flow through the laser. The box is filled with a mixture of eight parts helium, two parts CO_2 and two parts nitrogen. The gas has to be turned on in advance of any trial so that it has time to purge air out of the system, as air will cause arcing. H controlled the values of the gas cylinders, while I read out values on a flow meter in the gas line bolted to the side of the main frame of the laser.

The first trial

At 12.05 H had a cigarette preparatory to the first test of the laser. A small crowd assembled for the button-pressing and the usual nervous jokes were exchanged. After switching on, it was found that the power unit was dead because the cable that connected it to the mains was missing! This was soon rectified. The first proper trial proved that the spark gap was firing a visible flash but there was no sign of life in the laser box even when we closed the doors and turned off the light. H put the laser through a number of trials, increasing the voltage each time until there was a very loud bang indeed accompanying an arc discharge somewhere in the electronic components.

H summed up the results of this first set of trials:

> So what's wrong ... the trigger works, the spark gap seems to work, the charging of the first capacitor seems to work but it doesn't seem to get to the second one.

Further examination of the circuits revealed that both of the capacitors had been connected the wrong way round; these new capacitors had a preferred polarity. H was slightly troubled in case they might have been damaged by being charged incorrectly.

At 12.30 we tried the laser again. Once more the spark gap fired but there was no discharge in the laser. H tested the bottom capacitor by discharging it to earth with a wire. This produced a satisfactory spark showing that it was indeed holding charge. He then tried all the components of the electronics with an 'Avometer', a device that would check that resistances and connections were correct. Again summing up, H remarked that the spark gap seemed to be working and that the indicated capacitor voltage was falling to about half its predischarge value, which fitted in with the behaviour of Jumbo.

The next step was to check the polarity of the spark gap, and try it out with the polarity reversed; H was not sure whether the trigger power supply gave a positive or negative impulse and the manufacturers had been unable to tell him!

By 1.15 the polarity on the spark gap had been reversed and it was time for another trial. A sequence of trials was made at higher and higher voltage. At first the spark gap fired, but nothing else happened.

H increased the voltage even more but still nothing happened in the laser cavity. Eventually we went for lunch.

More trials

After lunch H left me to work with two graduate students. The first thing we did was to check out the characteristics of Jumbo. We found that we could see pre-ionization in Jumbo at voltages as low as 30,000, whereas we had seen nothing up to more than 40,000 volts in the new device. We also tried moving the electrodes of the new laser closer together, to no effect. Then we disconnected one terminal of the secondary capacitor, and by earthing it, generated a spark which proved that this capacitor too was being charged when the spark gap fired.

The next hypothesis was that the pre-ionization circuit was not working properly. We decided therefore to disconnect the pre-ionization circuit in Jumbo, to see what happened when that laser was disabled in that way. We found that the disabled Jumbo was completely quiescent up to at least 38,000 volts; this was consistent with the hypothesis that the pre-ionization was causing the trouble in the new laser.

Next, on H's return, we tried closing the electrode gap further, but the top electrode fell from its mounting and it was discovered that the minimum gap attainable was about 4 centimetres. First trials with the 4 centimetres configuration also produced no effect. We tried hard to reduce the light level so that even the minutest flicker would be visible but there was nothing to be seen. Eventually in some frustration H turned the voltage right up. There was a very very loud crack from an arc in the electronics and the spark firing mechanism blew up. This was rather depressing, and little more progress was made that day.

16 March

As we drove home I remarked to H that I had noticed that the trigger wires in the new laser were of much thinner gauge than those in Jumbo. He had not noticed this, but on the following morning confirmed it.

The plan for the following day was to exchange components one at a time between the new laser and Jumbo. Each component from the new laser would be tested in Jumbo, and if Jumbo still worked, then that part could be replaced in the new laser with confidence.

The first part to be exchanged was the rc circuit. After some trouble this was made to work perfectly in Jumbo. Next the new glass tubes containing their trigger wires were put into Jumbo. Before trying them H was quite confident that Jumbo would work with the new set in spite of their smaller diameter and different method of connection.

He was right; Jumbo did work with the new trigger wires, which were put back into the new laser.

Next, two of the leads between the capacitor and the top electrode were disconnected and Jumbo tried again. It worked without problem so the extra three leads were disconnected from the bottom electrode. Again Jumbo worked perfectly. It seemed that the number of connections and the distribution of charge were not problems and so it was assumed that the single connections on the new laser would be adequate.

We were now running out of things to test. We had tested the components in the laser box — that is, the single connection configuration, and the trigger wires — and we had tested the rc circuit. The new capacitors were much smaller but similar, and even better according to the manufacturer's specification. Nevertheless we were by now considering switching them to Jumbo, but because of the discrepancy in the physical size of the two sets, and the very short leads, this might prove difficult. Difficulties of a similar sort attended the mooted exchange of the bottom electrodes.

At this point, summing up the results achieved so far, H suggested that most of his suspicion hovered over the spark gap. He cited as evidence that the needle on the voltage indicator of the power supply had not been falling when the spark gap fired suggested that the primary capacitor was not discharging. There was some confusion here; earlier this needle had indicated the appropriate discharge.

We decided to try the new laser again before transferring the more difficult components. We would try working on the spark gap and firing it in new ways which bypassed the proper circuitry. This allowed us to test the laser without worrying about whether the circuit from the spark plug power supply, which included a dubious transformer (see note 6), was working. In this configuration that whole section of the circuit became redundant. Unfortunately, though the spark gap now worked, still nothing happened in the laser cavity even though the voltage was turned quite high.

Finally, short of anything else to do, at about 5:15 pm, H completely removed the connecting wire from the spark gap and tried again. Suddenly an arc discharge took place between the plates; a great cheer greeted it. Arc discharges continued as we increased the voltage and though we all felt relieved there was still no sign of pre-ionization and nothing that looked like a uniform discharge could be obtained. When I asked H to reconstruct the reasoning that led him to remove the trigger wire from the spark plug he commented:

> . . .so it seemed that as a final thing, if the trigger is not doing anything it might as well be disconnected — anyway why not do that, there may just be something that happens — that is obviously what must have been going

through my mind, so I just knocked it off, and ... hey presto ... if anything it was a bit of intuition.

Though we were pleased that we had achieved some kind of discharge between the electrodes of the new laser we had only now reached a stage that had been reached, in the case of Jumbo, in the middle of October 1974; Jumbo had not lased for a further five months! Celebration would have been premature, but we pressed on with the work for another hour that evening with an increased sense of optimism. We tried changing the gas mixture, and changing the electrode separation, but nothing but arc discharges could be obtained.

At one point I noticed that the anode of the new laser was heavily marked by the arc discharges. I pointed this out, but H shrugged it off. It now seemed that anode-marking was a normal part of these lasers' operation, so what had once been the 'vital clue' was now a piece of erroneous information!

We finally left the laboratory at about 6.30 pm and drove to H's house, where he thought about things that could be checked on Monday morning. He would check the value of the inductance and he could try swapping the capacitors even though this was a difficult task. He could also try checking the length of cables even where the inductances of these could not imaginably affect the performance of the laser. I had noticed another difference between the two lasers that I decided not to mention. This was that the bottom electrode of the new laser was about twice as thick as that on Jumbo. The two piece construction aside, H described the two electrodes as *identical* but I thought they were different. H summed up the rationale and the conclusions from two days' work as follows:

> the reason behind duplicating a Jumbo system was that I wanted lasers as quickly as possible — high powered lasers to be able to get on with a lot of the research which we've been trying to do with one laser and to duplicate Jumbo was definitely the right approach ...
> ...the fact that we've had two days' efforts, and we still haven't got a duplicate system to work is a reflection of kinds of problems you could face if you had tried to build another kind of laser system where I could be in the same mess as I was with Jumbo which may involve a year's work messing around getting the damned thing right.

I now left the scene. On the morning of Monday 19 March H went back to the laboratory to continue work. He reported events to me by telephone, in the middle of that same Monday afternoon. H had managed to get a uniform discharge in the laser box after not more than about an hour's work.

Lasers and knowledge

There is no need to labour further the point about the capricious nature of laser building skill and its transfer, but two more major points should be mentioned.

The first of these concerns Harrison's own developing tacit knowledge, its nature and limitations.

Throughout, Bob Harrison and I had been discussing similarities and differences between the old and the new laser. I had noticed the different thickness of the wires and had suggested that this might be significant. One of the graduate students had agreed that the thinner wires would have a significantly reduced surface area which might prevent proper pre-ionization. Yet Harrison had failed to see this as a significant difference; as it turned out his *not seeing* the difference was the proper way to see things. They were in fact just 'wires'. One might say that H's tacit knowledge was developed to an extent that enabled him to not to see a difference that was not a difference in terms of the operation of the laser. We others, without that level of ability, saw difference where it was inappropriate. Our suggestion that the difference between the wires might be important was a kind of laser builders' *faux pas*. It was as though we had suggested to a racing cyclist that his bike might go quicker if it were painted red instead of black.

There were other differences that I noticed and that Harrison ignored, quite rightly as it turned out. For example he was quite right not to see the 'obvious' physical differences in the bottom electrodes; I could see that the new one was much thicker than the old. He saw the bottom electrode as a 'double-discharge TEA-laser electrode' whereas I saw it as a piece of metal.

Again, the different ways the trigger wires were connected struck me as a possible cause of trouble. Conducting glue seemed very different to screw connections into a brass block, but Bob Harrison was right in not seeing this as a significant variation. Furthermore, he had learned to ignore the once crucial 'anode marking' and the exact degree of flatness of the glass tubes. Without knowing how to ignore all these things we might have spent months checking them out just as months were spent when Jumbo was being built. None of the things that Harrison had learned to ignore would be obviously significant, or insignificant, on a circuit diagram or in a technical article. The range of things to be ignored is, of course, indefinitely long.

On the other hand, in developing his laser building skill Harrison had also learned to see significance where previously he had noticed nothing. For example, the wires between the capacitors and the electrodes were no longer to be seen just as 'wires'; henceforward their length was crucial. One of the qualities of those particular wires was 'length'. This was a quality possessed by none of the other wires belonging to the laser (outside of the trigger wires themselves). All his previous electrical socialization had taught Harrison that wires did not have lengths. A long wire was 'the same' as a short wire; it is only in a few areas of electrical society — notably electronics — that wires do have lengths.

One might say that learning tacit knowledge, or acquiring culture, is a matter of learning this indefinitely long list of what is insignificant and, *inter alia*, learning what is significant. It entails learning that what might seem to the unskilled, or the uncultured, as going on in a different way is in fact going on in the same way and that what might seem to the uncultured as going on in the same way is in fact going on in a different way. It was in this that Harrison's new expertise lay.

Nevertheless Harrison's taken-for-granted reality, was not absolutely secure; it was not, after all, a formal system of knowledge. His hunches were better than mine but, as troubles developed, that is, as the laser continued to refuse to work, shadows of uncertainty began to creep in. All sorts of things were tested in spite of the fact that tacit knowledge would normally exclude them from consideration. Harrison's preparedness to test out some of my 'half-baked' suggestions was something more than good manners. What is more, as we drove home on the Friday night, some of the plans we hatched were a little desperate. Disorientation and bold speculation characterize many areas of cultural and political life when taken-for-granted reality is seriously disturbed.

When is something working properly?
The second major point to be drawn, especially from the Heriot-Watt study, is the difficulty of testing both knowledge and apparatus by means other than by having them perform the specific task toward which they are directed. Again, bicycle riding is an appropriate analogy; one only knows if one can ride by trying to ride. It should by now be clear that Harrison's laser building ability could only be tested by his trying to build a laser; the same principle applies to specific parts of the apparatus.

For example, at one time we tried to measure the performance of the spark-firing transformer. This needed a complex network of measuring tools since the voltages involved were so high. Our initial conclusion was that the transformer was misbehaving because it seemed to be stepping up the input voltage by a factor of ten rather than the specified factor of one hundred. But we did not fully believe this. We suspected that there was some flaw in the network of testing instruments we had used. We were thrown back upon trying the transformer in the laser as a test of its functioning![6]

Most of the other tests we had done would have been immensely time consuming if there were no Jumbo to use as a facsimile test bed for the new laser. If we had been working with the new laser alone, it would have been impossible to be certain about which components, or even how many components, were broken down at any one time. It is possible to test certain components for their specified performance where this is a simple quantity; for instance, components can be tested for their electrical resistance with an 'Avometer' and similarly, connections can be tested by making certain that they have nil resistance and isolation can be tested by looking for infinite resistance. Other components can be checked by metering their output. The difficulty is that where components are being used in new ways — as in any new piece of apparatus — they will not necessarily perform in the desired manner even though they appear to meet specification. For instance, it will be remembered that Jumbo had troubles with arc discharges from one earth point to another during construction. Any simple meter test would see nothing but nil resistance between two such points. Consider also that the length of the leads between the capacitors and the top electrode was critical. Only the most sophisticated test could reveal the difference in inductance between eight-inch high tension leads and six-inch high tension leads, yet it was a difference of this order that was claimed to make a significant difference to the laser performance.

It would, of course, be possible to devise tests for such differences, but the tests would be of such complexity — they would have to test components' performance in the face of high tension surges of the same potential and time profile as used in the laser — that the apparatus for testing would come to resemble the laser itself. And, of course, the exact specification of the pulse profile of the laser was not known, and would have been very difficult to measure. (Remember the failure to measure the pulse profile of Jumbo using a probe coil.)

Similarly, it is almost impossible to think of a way the bottom electrode trigger wire assembly could have been tested without either fitting it to Jumbo, or making the new laser work. The same applies to the question of the number of connections to the electrodes.

H put some of these problems this way:

> . . . really to set something up under the same experimental conditions that you have from your laser is pretty difficult.
> . . . You always come away thinking 'well maybe that wasn't a completely conclusive test'.

Thus it seems fair to say that the operating laser *defined* the experimental conditions under which the components of the laser had to work. The desired component specification was embodied in the laser rather than in a set of performance figures.

The laser studies: five propositions

The two detailed examinations of the process of laser building make the findings of the knowledge transfer network study quite comprehensible. For example, it is easy to see that a laser builder might fail completely to make his laser work even where his knowledge sources were good. Harrison failed to make his first laser work for several months, even though it seemed to be a perfectly good copy of one that was functioning elsewhere. One can also see that it is important that one's source of knowledge is a competent laser builder. Harrison would not have been a lot of use as an informant at the beginning of his attempt to build Jumbo; there is no way that he could have informed anyone about the necessity of having the leads from the capacitor to the electrodes as short as possible, for example, since he did not realize the importance of this himself. *But,* he did not know that he did not know. These points can be represented as two propositions:

Proposition One: Transfer of skill-like knowledge is capricious.
Proposition Two: Skill-like knowledge travels best (or only) through accomplished practitioners.

Another thing to be noted is that it took Bob Harrison six months to make his first laser work from the moment of first test, whereas it took him only two days for his second device. Successful laser builders, then, are 'competent practitioners'. This means more than that they have built a laser. It means that they possess new skills and new knowledge which enables them to build another one faster. Working with H the second time, this became clear in the way he confidently ignored certain parameters that were once thought vital (flatness of tubes, anode marking) and in the way he overlooked differences between the lasers that I noticed and would have thought important: for example, the thickness of the bottom electrode, the glue and the wires. It was even evident in certain displays of more ordinary perceptual skill such as his ability to hear the quality of the sound made by an arc discharge and thereby know the discharge's characteristics. At the same time the fact that it did take H two whole days to sort out the problems and that a lot of what was done was a question of trial and error shows that H's new abilities did not lie (or at least not entirely) in new information.

Proposition Three: Experimental ability has the character of a skill that can be acquired and developed with practice. Like a skill, it cannot be fully explicated or absolutely established.

From the three studies it seems firmly established that laser-building ability is something you do not know whether you possess until you have built a laser. Thus, laser building knowledge is invisible in its passage. There is no way, such as by examining the available range of information sources, or by making an inventory of items of information known, that will reveal whether a scientist has laser building ability.

> *Proposition Four:* Experimental ability is invisible in its passage and in those who possess it.

A closely related point is that the only indicator that someone has laser-building ability is his or her ability to build a laser. The correct functioning of the apparatus and the experimenter is defined by the output of the apparatus. This can be seen very nicely in the third study in respect of the various parts of the new laser; what counted as working properly was defined by ability to function first in Jumbo. That is, parts were defined as good parts, irrespective of other measurements, if they could take part in the process of lasing. It is important to note that it was extremely difficult, if not impossible, to think of any other way to discover their nature. Even where technical means appeared to be available, measurements were distrusted as surrogates for testing within the laser. Thus:

> *Proposition Five:* Proper working of the apparatus, parts of the apparatus *and the experimenter* are defined by the ability to take part in producing the proper experimental outcome. Other indicators cannot be found.

Crystallization of certainty

Finally, a more subtle conclusion may be drawn. This is that scientists are resistant to the sort of account of experimentation that I have just given. For example, it is tempting to think that if H had not been so stupid as to reverse the polarity in Jumbo, he might have had it working quite quickly. He himself was inclined to be 'wise after the event' and blame himself for his own stupidity. He felt it was a simple case of human error rather than a failure of skill.

The same applied to his eventual solution of the problems of the Heriot-Watt laser: The first thing Bob Harrison did on the Monday morning was to run both lasers on pure helium to see if this would make the discharge easier to obtain. He discovered that even Jumbo would only arc under these circumstances. Then, in adjusting the gas flows back to their proper values in the new laser, he noticed that the flow meter valve seemed to be hypersensitive; further examination revealed that the flow meter assembly was designed to pass only one litre of gas per minute rather than ten litres per minute. He changed this assembly, left the laser to flush with gas at the appropriate rate for a half hour, and then keeping the spark gap in the same configuration as we had used on the previous Friday evening, he found that he obtained a uniform discharge at 42,000 volts — consistent with the behaviour of Jumbo. After this, lasing is easily obtained by assembling the mirrors, and so forth. H reasoned that we had not been able to obtain a uniform discharge before because the gas flow had been so slow that the laser box had never been properly flushed of its residual air; this is what he now believed had been the cause of the continual arcing.

Bob Harrison reconstructed what had happened:

> The trouble with the system has been one of human error . . . Except for the trigger unit — that would have blown up anyway. Everything else was the fault of the experimenter. You always find this. There are too many things to think about, etc. There's a limit to how much checking of the system you are prepared to do before trying it out. You make your own assessment of the situation and draw the conclusion that at that stage you can learn a lot more about your system from trying it out rather than from over-checking before trying it out. Obviously if I'd been really thorough I'd have spotted the flow meter problem before trying it out.

I believe that H is being rather hard on himself here. There is no reason to suppose that even if the gas mixture had been correct the laser would have worked before 5.15 pm on the previous Friday when the first arc discharge was obtained in the laser cavity. Before then, so it seems, all the charge had been leaking away down the spark plug wire. But, with the laser working, the uncertainty which surrounded the laser on the Friday evening suddenly crystallized out. Physics and laser technology resumed their familiar sharp outline, so that the failures of the previous two days were now seen as a consequence of human error disturbing natural regularity. It is human to err, but we do not think that it is natural to err.

One must distance oneself from the standard view of experimentation in science and escape from the railroad of common sense to see the conventional nature of this reconstruction of 'what really went on' in an experiment. I reserved the description of the solution of the Heriot-Watt laser problem to this late stage so that the reader could avoid being wise after the event for as long as possible.

The standard view of experiment is very much in accord with the view of the information scientist. Thus, once a scientist has mastered the basic skills of his trade he or she ought to be able to repeat any experiment using just what information is available from the usual sources. This is the initial 'murine' view as outlined by Popper and Dennis in Chapter Two. This view rests upon the notion that scientific facts are *testable* by 'independent replication'. The notion of independent testability ignores the active part played by man in seeing regularity rather than passively registering it. The point of Popper's third quotation is evident in the contrasting similarities and differences that Harrison and I saw in the two big lasers. The 'correct' way of seeing was only finally established after the laser worked. Prior instruction could not encapsulate these ways of seeing.

But, as soon as an experiment is successful — which in turn reaffirms the independent regularity of nature — any irregularity must be explained as arising out of human error. One experiences this sudden switch in perception in many practical activities. It is, as I have suggested, like a process of crystallization. One moment nature is obscure and recalcitrant, the next moment everything works and nature is once more orderly. The earlier obscurity and recalcitrance, which demanded so much human intervention to regulate, is then displayed as a *defect* in the human contribution.

I am certain that any working scientist will recognize all the elements in these stories from his or her own experience. And yet, it is so hard to maintain a degree of consciousness of such familiar happenings that it is almost impossible to see that they add up to a consistent story. Bob Harrison, as we noted, switched back in an instant to his picture of nature as orderly and cooperatively passive. This provides a sixth proposition:

> *Proposition Six:* Scientists and others tend to believe in the responsiveness of nature to manipulations directed by sets of algorithm-like instructions. This gives the impression that carrying out experiments is, literally, a formality. This belief, though it may occasionally be suspended at times of difficulty, re-crystallizes catastrophically upon the successful completion of an experiment.

It is this crystallization and re-crystallization that helps maintain existing scientific knowledge. Doubts, if they arise, last for only a very short time. In the chapters that follow we will explore the significance of Propositions One to Six for areas of less straightforward science.

Notes

1. There now follow three case studies. Case studies must be 'representative' if the conclusions drawn are to be generalized. The generalized conclusions of these case studies will be applied to science as a whole, and then to culture as a whole. For justification of the suitability of these studies and for a description of the fieldwork see the methodological appendix.

2. The cost of the equipment might be between £500 and £2,000, mostly for mirrors and ancillary equipment such as oscilloscopes and detectors which would be found in any laser laboratory.

3. In 1971, at the outset of these studies on lasers, the project was not envisaged as a study of replication but of knowledge transfer. For further discussion, see the methodological appendix.

4. The laboratories were found by a snowball technique: I asked at each location for the names of others making TEA-lasers. These laboratories were visited and one or more of the scientists involved centrally in the construction of the laser was interviewed at each site.

5. Polanyi writes as follows about tacit knowledge:

> Science is operated by the skill of the scientist and it is through the exercise of his skill that he shapes his scientific knowledge.
>
> ... the aim of a skilful performance is achieved by the observance of a set of rules which are not known as such to the person following them.
>
> ... from my interrogations of physicists, engineers and bicycle manufacturers, I have come to the conclusion that the principle by which the cyclist keeps his balance is not generally known. [and much more]
>
> An art which cannot be specified in detail cannot be transmitted by prescription, since no prescription for it exists. It can be passed on only by example from master to apprentice. This restricts the range of diffusion to that of personal contacts. (1958, pp 52-3)

There is some danger in completely identifying Wittgenstein's ideas and phenomenological ideas with Polanyi's tacit knowledge even though the term is useful and the consequences are similar. The formulation can be misleading for it tends to suggest that the only reason that knowledge cannot be formalized is because there is something *hidden*. Tacit knowledge could be converted into information, it seems to suggest; it is only time and ignorance that prevents us doing so. While it is true that the development of sciences does seem to involve a degree of explication of what was once only vaguely apprehended, the underlying model, it must be remembered, is the 'form of life' and it is incorrect to think that it could be eliminated if enough determination were put into the task.

A second danger is that Polanyi seems to take the idea of tacit knowledge much further than we would want to. For example, he believes that the solutions to scientific problems are somehow anticipated by scientists by virtue of their tacit knowledge. There may be some truth in this, evident in the way we solve chess problems and the like, by virtue of our ability to comprehend more of the context of a problem than we can articulate, but Polanyi's phrasing seems to suggest still more (eg see his 1966, pp 21-2).

The reason for continuing to use the term tacit knowledge, in spite of its undesirable connotations, is that there is no other way of referring to what we know by virtue of participation in a form of life, nor what is learned as one moves from being a non-participant to a participant. The Wittgensteinian model, as well as the phenomenological model, is set in an unchanging, undeveloping world (see Chapter One).

Ravetz (1971) draws on Polanyi to stress the craft aspect of scientific work. He describes scientific activity as having a 'peculiar' character 'as a special sort of craft work operating in intellectually constructed objects' (p 146). This leads him to stress the craft component in scientific method, the universality of pitfalls, the uncertain nature of criteria of adequacy in scientific assessment, and the interpersonal nature of some components of scientific communication.

The difficulty is that Ravetz' determination to treat the findings of science as 'objective' tends to obscure the sociological significance of his scholarly work. His argument is frequently qualified in surprising and seemingly inconsistent ways (eg, see p 178 and p 147).

6. No oscilloscope in the department could handle 35,000 volts so it was decided to test the output of the transformer by using a 'high voltage probe' that should step the potential down again by a factor of a thousand. The net result of a time consuming test, involving several pieces of apparatus (Jumbo's spark power supply, the transformer, the high voltage probe, and the oscilloscope), was the less than fully credible conclusion that the transformer was stepping up by only about ten times rather than a hundred.

To test our own measuring process, H decided to test an identical transformer in the same way. This produced the same result! Thus it looked as though the measuring process had gone wrong somewhere by a factor of ten, but we could not see how. The other alternative was that both transformers were wrong by a factor of ten when supplied by the manufacturer but this manufacturer only made this one type of transformer, so it was unlikely that units of the wrong value could have been supplied. At my suggestion, the transformers were tested again at a much lower input voltage by providing a few volts from a small pulse generator. On this test they seemed to work consistently and with a hundred-fold increase in potential. Then we tried to test the output of Jumbo's transformer, in order to test our initial measurements. Here however, we ran into great problems because we could not tap the potential without arcing problems. In the end, this whole series of tests was inconclusive and we were left uncertain whether the supplied transformers both worked properly at low voltage, but not at high voltage, or whether there was something wrong with our measuring techniques, or some piece of the measuring apparatus such as the high voltage probe. Only a test on the laser, with everything else working, would be completely decisive. Unfortunately, in this case, the transformer could not be fitted to Jumbo because the parameters of Jumbo's spark pulse generator and spark gap were different.

Chapter Four

Detecting Gravitational Radiation:
The Experimenters' Regress

None of the scientists I spoke to thought that the TEA-laser was especially difficult to make. Though there are references to the 'black art of radar electronics', no one suggested that I was looking at an untypical experimental area. Nor does there seem to be anything intrinsic about a TEA-laser that would make working with it different from working with other experimental apparatus. It is an ordinary sized, ordinary-priced piece of laboratory apparatus using no especially exotic bits of technology or rare materials. It doesn't require special conditions of cleanliness or sterility. It is tolerant to kicks, bangs and having foreign bodies left in its workings. (A bolt and a screwdriver were left in the gas tubes of Jumbo and the Heriot-Watt laser, respectively.) The very first printed information about the laser stressed its robustness by calling it a 'plywood laser'.

If the laser is untypical, it is because it works with higher voltages than are usually encountered; however, this in turn demands that all its parts are large, separate and readily visible. There is nothing 'fiddly' involved; there are no extremes of temperature or pressure, nor does the device require isolation from its surroundings either electrically, acoustically, magnetically or seismically. Laser experimentation is very ordinary and very easy, especially compared with the experiments which are to be described.

Gravitational radiation: 1972
Gravitational radiation can be thought of as the invisible gravitational equivalent of light or other electromagnetic radiation (see, for example, Davies, 1980). Most scientists agree that Einstein's general theory predicts that moving massive bodies will produce gravity waves: however, they are so weak that their detection is very difficult. For example, no one has so far suggested a way of generating detectable fluxes of gravitational radiation on Earth, at least not within the foreseeable future. Nevertheless, it is now accepted that some sensible proportion of the vast amounts of energy generated in the violent events of the universe should be dissipated in the form of gravitational radiation which may be detectable on Earth. Exploding supernovae, black holes and binary stars should produce sizeable fluxes of gravity waves which would show themselves on Earth as a tiny oscillation in the value of 'G' — the constant that is related to the gravitational pull of one object on another.

Just as the planets are attracted to the Sun and to one another by the force of gravity, so are smaller objects. We know that the Earth's gravitational pull is strong enough to keep us firmly anchored to the ground most of the time, but we are also attracted to one another by gravitational forces. We don't stick together because the forces are almost immeasurably small. It was a triumph of experimental science when, in 1798, Cavendish measured the gravitational attraction between two massive lead balls. The attraction between them comprised only one five hundred millionth of their mass! Looking for gravitational radiation is unimaginably more difficult than looking for this tiny force because the effect of a gravity wave pulse is no more than a minute fluctuation within it. For example, one of the smaller antennae (the detectors are often referred to as antennae) that I was shown was encased in a glass vacuum vessel. The core consisted of, perhaps, one hundred kilograms of metal, yet the impact of the *light* from a small flashgun on the mass of metal was enough to send the recording trace off the measuring scale. And this was a fairly insensitive test for one of these devices.

Design of a gravity wave detector

This, then , was a difficult experiment. The standard technique was developed by Professor Joseph Weber (pronounced 'Whebber') of the University of Maryland. He looked for changes in the length (strains) of a massive aluminium alloy bar caused by the changes in gravitational attraction between its parts. Such a bar, often weighing several tons, could not be expected to change its dimensions by more than a fraction of the radius of an electron as a pulse of gravitational radiation passed. Fortunately, the radiation is an oscillation and, if the dimensions of the bar are just right, it will vibrate, or 'ring' like a bell, at the same frequency as the radiation. This means that the energy in the pulse can be effectively integrated or aggregated into something just barely measurable.

A Weber-bar detector, or antenna, comprises the heavy bar with some means of measuring its vibrations. Most designs used strain sensitive 'piezo-electric' crystals glued, or otherwise fixed, to the bar. These crystals produce electricity when they are deformed. In a gravity wave detector the voltage produced is so small as to be almost undetectable. Thus, a critical part of the design is the signal amplifier. Once amplified the signals can be recorded on a chart recorder, or fed into a computer for immediate analysis.

Such devices, of course, cannot distinguish between vibrations due to gravitational radiation and those induced by any other force. Thus, to make a reasonable attempt to detect gravity waves, the bar must be insulated from all other known and potential disturbances such as electrical, magnetic, thermal, acoustic and seismic forces. Weber attempted to do this by suspending the bar in a metal vacuum chamber on a thin wire. The suspension was insulated from the ground by a series of lead and rubber sheets. (The seismic insulation seems to have been a particularly simple and ingenious solution to what many had thought to be an insoluble problem.)

In spite of these precautions, the bar will not normally be completely quiescent. So long as it is at a temperature above absolute zero, vibrations will be induced in it by the random movements of its own atoms. Thus, the strain gauges will register a continual output of thermal 'noise'. If this were recorded on graph paper by a pen recorder (as it was in many experiments), what will be seen is a spiky wavy line showing random peaks and troughs. A gravity wave would be represented (perhaps) as a particularly high peak, but a decision has to be made about the threshold above which a peak counts as a gravity wave rather than noise. However high the threshold, it must be expected that occasionally a peak due entirely to noise would rise above it. In order to be confident that some gravity waves are being detected, it is necessary to estimate the number of 'accidental' peaks one should obtain as a result of noise alone, then make certain that the total number of above-threshold peaks is still greater. (See technical appendix at end of the chapter for more details of the process of gravity wave detection.) In 1969 Weber claimed to have detected several (about seven) peaks every day which could not be accounted for by noise in the detector.

Status of Weber's claims

Weber's claims are now nearly universally disbelieved. Nevertheless, the search for gravitational radiation goes on, and many current experimental devices are similar to Weber's. Weber's findings were sceptically received because he seemed to find far too much gravitational radiation to be compatible with contemporary cosmological theories. The apparatus now under development to detect fluxes of radiation in line with cosmologists' predictions is meant to be 10^9 (one thousand million) times more sensitive. Thus, though I am going to talk of the extinction of certain claims to have found a new natural phenomenon — gravity waves — it must be understood that I refer only to the phenomenon claimed to have been discovered by Weber — high fluxes of gravity waves.

Weber's detection rate seemed far too great when calculations of the probable sensitivity of his antenna were compared with the amounts of energy, dissipated in the form of gravity waves, that should be generated by cosmic events. If Weber's results were extrapolated, assuming an isotropic (uniform) universe, and assuming that gravitational radiation was not concentrated into the

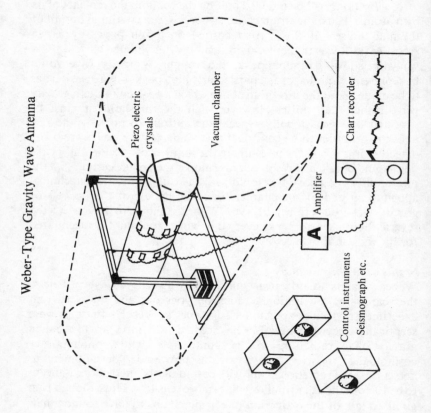

Weber-Type Gravity Wave Antenna

Piezo electric crystals

Vacuum chamber

A   Amplifier

Chart recorder

Control instruments
Seismograph etc.

1661 Hertz (cycles per second) frequency that Weber could best detect, then the amount of energy that was being generated in the cosmos would imply an unreasonably short lifetime. The universe must soon be completely 'burned up' if it were to continue to radiate in this way. These calculations suggested that Weber must be wrong by many orders of magnitude.

Though Weber's first claims were not entirely credible, in the early years of the 1970s he produced a series of ingenious modifications which led other laboratories to attempt to replicate his work. One of the most important new pieces of evidence was that above-threshold peaks could be detected simultaneously on two or more detectors separated by a thousand miles. At first sight, it seemed that only some extraterrestrial disturbance, such as gravity waves, could be responsible for these simultaneous observations. Another piece of evidence was that Weber discovered a periodicity in the disturbances of around twenty-four hours. This suggested that the radiation came from one extraterrestrial direction only. What is more, the periodicity seemed to relate to the Earth's disposition with regard to our galaxy, rather than with regard to the Sun, and this suggested an extra-solar (or galactic) source for the disturbance. (This effect became known as the 'sidereal correlation'; see technical appendix.)

By 1972, at the time of the fieldwork now to be discussed, several other laboratories had built or were building antennae to search for gravitational radiation. Apart from Weber, three others had been operating long enough to be ready to make their own claims. All of these claims were negative.

The experimenters' regress
So far I have described the general principles of the detection of gravitational radiation and a few of its problems. The naïve, but scientifically accomplished, reader might feel that, given some time, he or she now knows how to build a gravity wave detector. What is needed is a vacuum chamber; a heavy bar of aluminium alloy to be suspended within it and insulated from magnetic and electrical forces and from the ground with a lead and rubber pile. Piezo-electric strain gauges must be attached to the bar and their signals amplified and recorded. The whole device can be built in a year or two at the cost of less than £50,000.

But now we must remind ourselves of Propositions One to Six at the end of the previous chapter. Proposition One to Proposition Four show us that it is most unlikely that we could now build a gravity wave detector. Proposition Six suggests why we might have been fooled into thinking that we now had the recipe for detecting gravity waves. Proposition Five suggests that we will have no idea whether we can do it until we try to see if we obtain the correct outcome. *But what is the correct outcome?*

What the correct outcome is depends upon whether there are gravity waves hitting the Earth in detectable fluxes. To find this out we must build a good gravity wave detector and have a look. But we won't know if we have built a good detector until we have tried it and obtained the correct outcome! But we don't know what the correct outcome is until . . . and so on *ad infinitum*.

The existence of this circle, which I call the 'experimenters' regress', comprises the central argument of this book. Experimental work can only be used as a *test* if some way is found to break into the circle. The experimenters' regress did not make itself apparent in the last chapter because in the case of the TEA-laser the circle was readily broken. The ability of the laser to vaporize concrete, or whatever, comprised a universally agreed criterion of experiment quality. There was never any doubt that the laser ought to be able to work and never any doubt about when one was working and when it was not. Where such a clear criterion is not available, the experimenters' regress can only be avoided by finding some other means of defining the quality of an experiment; a criterion must be found which is independent of the output of the experiment itself.[1]

Scientists at their work

If the experimenters' regress is a real problem we might expect to find scientists disagreeing about whether their own and others' devices were good detectors. They are likely to do this in the absence of an independent criterion for determining quality. This is precisely what was found in my fieldwork and it can be illustrated with quotations from interviews with the scientists involved. The following sets of comments, taken from interviews conducted in 1972, reveal differences in scientists' assessments of the value of experiments.

The first set of comments shows the variation in scientists' opinions regarding the value of others' experimental set-ups and reported results. In each case, three scientists at different establishments are reporting on the experiment of a fourth.

Experiment W

Scientist a: . . . that's why the W thing, though it's very complicated, has certain attributes so that if they see something, it's a little more believable . . . They've really put some thought into it . . .
Scientist b: They hope to get very high sensitivity but I don't believe them frankly. There are more subtle ways round it than brute force . . .
Scientist c: I think that the group at . . . W . . . are just out of their minds.

Experiment X

Scientist i: ...he is at a very small place ... [but] ...I have looked at his data, and he certainly has some interesting data.
Scientist ii: I am not really impressed with his experimental capabilities so I would question anything he has done more than I would question other people's.
Scientist iii: That experiment is a bunch of shit!

Experiment Y

Scientist 1: Y's results do seem quite impressive. They are sort of very business-like and look quite authoritative....
Scientist 2: My best estimate of his sensitivity, and he and I are good friends ... is ... [low] ... and he has just got no chance [of detecting gravity waves].
Scientist 3: If you do as Y has done and you just give your figures to some ... girls and ask them to work that out, well, you don't know anything. You don't know whether those girls were talking to their boyfriends at the time.

Experiment Z

Scientist I: Z's experiment is quite interesting, and shouldn't be ruled out just because the ... group can't repeat it.
Scientist II: I am very unimpressed with the Z affair.
Scientist III: Then there's Z. Now the Z thing is an out and out fraud!

The second set of comments shows that scientists perceived the importance of minor variations in bar-type detectors differently; they had different perceptions of what were to be counted as copies of which others. They saw similarity and difference in different ways.

i: You can pick up a good text book and it will tell you how to build a gravity wave detector .... At least based on the theory that we have now. Looking at someone else's apparatus is a waste of time anyway. Basically it's all nineteenth-century technology and could all have been done a hundred years ago except for some odds and ends. The theory is no different from electromagnetic radiation ...

From this point of view, all the detectors built should be capable of seeing the radiation if it is there, for there are no particular problems associated with building them.

ii: The thing that really puzzles me is that apart from the split bar antenna (the distinctive British version of the device) everybody else is just doing carbon copies. That's the really disappointing thing. Nobody's really doing research, they're just being copy cats. I thought the scientific community was hotter than that.

This scientist (whose apparatus was least like the originator's) perceived all the others as though they *were* carbon copies. To him the differences between the detectors were not significant.

On the other hand, the following two remarks show that the differences rather than the similarities between detectors may be perceived as more significant in respect to seeing the radiation.

iii: . . . it's very difficult to make a carbon copy. You can make a near one, but if it turns out that what's critical is the way he glued his transducers, and he forgets to tell you that the technician always puts a copy of *Physical Review* on top of them for weight, well, it could make all the difference.

iv: Inevitably in an experiment like this there are going to be a lot of negative results when people first go on the air because the effect is that small, any small difference in the apparatus can make a big difference in the observations. . . . I mean when you build an experiment there are lots of things about experiments that are not communicated in articles and so on. There are so called standard techniques but those techniques, it may be necessary to do them in a certain way.

Finally, Weber saw the dissimilarities between the detectors as the dominant feature, and felt that these differences made the secondary detectors less effective than his own device.

v: Well, I think it is very unfortunate because I did these experiments and I published all relevant information on the technology, and it seemed to me that one other person should repeat my experiments with my technology, and then having done it as well as I could do it they should do it better . . . It is an international disgrace that the experiment hasn't been repeated by anyone with that sensitivity.

The third set of evidence reveals variations in scientists' perceptions of the value of various parts of the originator's experimental procedures. Weber had been reporting results and had been heavily criticized for some time before the majority of secondary scientists began setting up their own experiments. In answering criticisms and improving his results Weber produced a series of experimental elaborations one or other of which convinced different secondary scientists to take the findings seriously enough to start work themselves. The first of these elaborations was the demonstration of coincident signals from two or more detectors separated by large distances. Some scientists found this convincing. Thus one scientist said:

[ ] wrote to him specifically asking about quadruple and triple coincidences because this to me is the chief criterion. The chances of three detectors or four detectors going off together is very remote.

On the other hand, some scientists believed that the coincidences could quite easily be produced by electronics, chance, or some other artefact.

. . . from talking it turns out that the bar in [ ] and the bar in [ ] didn't have independent electronics at all. . . . There was some very important common content to both signals. I said . . . no wonder you see coincidences. So all in all I wrote the whole thing off again.

However, Weber then ran an experiment where the signal from one of the bars was passed through a time delay; he showed that under these circumstances the coincidences disappeared. This, of course, suggested that they were not artefacts of electronics or chance. Several respondents made remarks such as '. . . the time delay experiment is very convincing', whereas others did not find it so.

Weber's discovery of the correlation of peaks in gravity wave activity with sidereal time was the outstanding fact requiring explanation for some scientists:

> . . . I couldn't care less about the delay line experiment. You could invent other mechanisms which would cause the coincidences to go away. . . . The sidereal correlation to me is the only thing of that whole bunch of stuff that makes me stand up and worry about it. . . . If that sidereal correlation disappears you can take that whole . . . experiment and stuff it someplace.

Against this, two scientists remarked:

> The thing that finally convinced a lot of us . . . was when he reported that a computer had analysed his data and found the same thing.
> The most convincing thing is that he has put it in a computer . . .

But another said:

> You know he's claimed to have people write computer programs for him 'hands off'. I don't know what that means. . . . One thing that me and a lot of people are unhappy about, is the way he's analysed the data, and the fact that he's done it in a computer doesn't make that much difference. . . .

Fourthly, the list of 'non-scientific' reasons that scientists offered for their belief or disbelief in the results of Weber's and others' work reveals the lack of an 'objective' criterion of excellence. This list comprised:

> Faith in experimental capabilities and honesty, based on a previous working partnership.
> Personality and intelligence of experimenters.
> Reputation of running a huge lab.
> Whether the scientist worked in industry or academia.
> Previous history of failures.
> 'Inside information'.
> Style and presentation of results.
> Psychological approach to experiment.
> Size and prestige of university of origin.
> Integration into various scientific networks.
> Nationality.

As one scientist put it, in outlining the source of his disbelief in Weber's results:

You see, all this has very little to do with science. In the end we're going to get down to his experiment and you'll find that I can't pick it apart as carefully as I'd like.

There is, then, no set of 'scientific' criteria which can establish the validity of findings in this field. The experimenters' regress leads scientists to reach for other criteria of quality.

Fifth, and finally, the following set of interview extracts shows that scientists are engaged in other than formal methods of argument and persuasion. In the first place, the following quotations should be read as showing lack of consensus over formal criteria.

We had a . . . summer school in . . . which lasted for two weeks. When we first went there, there was a certain amount of excitement in the air because it appeared from the discussion that 'P' was saying that he had looked for thirty days and had not seen any coincidences at the level of sensitivity of 'O' and 'O' was not feeling very good about it. They had many hours of talk together, and officially when the conference left, they agreed that 'O' was more sensitive than 'P' . . . So officially there wasn't any disagreement. Unofficially I don't know. . . .

Well there are two professors from . . . going round spreading rumours that 'Q' is repeating my experiment with one degree of magnitude greater sensitivity. This he has denied in a telephone conversation with 'R'. Where the truth lies, I don't know.

. . . when I heard about these results and the fact that this second-hand report came back that discredited 'O', I called up 'O' and told him what the rumours were that were going around so that he would know, and I said that I doubted if that was really the true picture, and he told me quite a lot more, and that they really didn't have anything near his sensitivity . . .

Then there was this rumour that 'S' had seen 'T's paper and decided that the statistics were absolute junk, and refused to publish it, and 'T' wanted him to publish it and he wouldn't publish it, and then when 'S' went on vacation, the other editor 'V' let it go past because he decided it was making too much noise.

. . . in fact there is a *samizdat* kind of thing circulating around which 'S' wrote which he didn't want to publish pointing out all these kind of inconsistencies.

There was an attempt by 'U' and a few others to get me involved and intrigued into a 'Young Turks' rebellion . . . I did not get involved in that. . . . 'U' is a negativist in his attitude toward things quite often, and he set out to topple 'O' for a while . . . There are sort of two camps and 'U' is the main activist of the camp that is out to topple 'O'.

We need go no further to see that, in this case, the resolution of the arguments about whose detectors were good ones and whose were bad was an interesting social process.

Competence and existence

The resolution of these arguments is coextensive with the question of whether gravity waves exist. When it is decided which are the good experiments, it becomes clear whether those which have detected gravity waves or those which have not are the good ones. Thus, whether gravity waves are there to be detected becomes known.

On the other hand, once it is known whether they are there to be detected, there is a criterion available to determine whether any particular apparatus is a good one. If there are gravity waves a good apparatus is one which detects them. If there are no gravity waves the good experiments are those which have not detected them.

Thus, the definition of what counts as a good gravity wave detector, and the resolution of the question of whether gravity waves exist, are congruent social processes. They are the social embodiment of the experimenters' regress.

The case of gravitational radiation leads to two more Propositions.

Proposition Seven: When the normal criterion — successful outcome — is not available, scientists disagree about which experiments are competently done.

Proposition Eight: Where there is disagreement about what counts as a competently performed experiment, the ensuing debate is coextensive with the debate about what the proper outcome of the experiment is. The closure of debate about the meaning of competence is the 'discovery' or 'non-discovery' of a new phenomenon.

Gravitational radiation: 1975

After 1972, events favoured Weber's claims less and less. In July 1973 negative results were published by two separate groups (one two weeks after the other) in *Physical Review Letters*. In December 1973, a third group published negative results in *Nature*. Further articles claiming negative results at increased sensitivity were subsequently published by these groups and also by three other groups. No one has since concluded that they found anything that would corroborate Weber's findings.

After 1972 the thrust of experimental activity changed along with the growth of certainty that Weber's results were incorrect. Whereas in 1972 about a dozen groups were engaged in active experimentation directed at Weber's findings, by 1975 no one but Weber was still working in this direction, and even he faced severe funding problems. However, about seven groups were building, or considering the design of, antennae of several orders of magnitude greater sensitivity in the hope of detecting the small, theoretically predicted, radiation flux.

In 1972, a few scientists believed in the existence of high fluxes of gravity waves, and hardly any would openly commit themselves to their non-existence. By 1975, a number of scientists had spent time and effort actively prosecuting the case against Weber. Most of the others accepted that he was wrong and only one scientist other than Weber thought the search for high fluxes worth pursuing. It might be fair to refer to 1975 as, in the words of one of my respondents, belonging to 'the post-Weber era'.

Closure of the debate
The details of the next phase of arguments concerning the existence of gravity waves — the arguments that led to the virtual extinction of the credibility of the high flux claim — have been described at length elsewhere (Collins, 1981c). Here I will give only a brief summary.

While nearly all scientists agreed by 1975 that the fluxes did not exist and that Weber's experiment was not adequate, their reasons differed markedly. Some had become convinced because Weber had made a rather glaring error in his computer programme at one point; others thought that the error had been satisfactorily corrected before too much damage was done. Some thought that the statistical analyses of the level of background noise and the number of residual peaks was inadequate, but others did not think this was a decisive point. Weber had also made a grave mistake when he claimed to have found coincident signals between his own detector and that of an entirely independent laboratory. These coincidences were extracted from the data by comparing sections of tape from the two detectors. Unfortunately for Weber, it turned out that the two sections of tape he compared had been recorded more than four hours apart so that he was effectively conjuring a signal out of what should have been pure noise. Once more, though, it was not hard to find scientists who thought that the damage had not been too great since the level of signal reported was scarcely statistically significant.[2]

Another factor considered important by some was that Weber did not manage to increase the signal-to-noise ratio of his results over the years. In fact, considering that his apparatus was undergoing continual improvement, the net signal seemed to be going down. This was not felt to be typical of new scientific work. What is more, the initially reported sidereal correlation faded away. Again, however, these criticisms were thought to be decisive by only one or two scientists.

Finally, it almost goes without saying that the almost uniformly negative results of other laboratories were an important point. However, all but one of the roughly (when is an experiment not an experiment?) six negative experiments were trenchantly criticized by one or more of Weber's critics! This is not to mention Weber himself who saw all six as inadequate. This should come as no surprise given the analysis in earlier sections.[3] Only one experiment remained immune to criticism by Weber's critics and this was an experiment designed to be as near as possible a carbon-copy of the original Weber design. Weber's criticism of it turned on differences in the signal processing algorithm (Collins, 1981c).

Thus, one or more of his critics, along with Weber himself, found fault with every one of the arguments and pieces of evidence directed against the high flux claim. Only in the case of one experiment was Weber without an ally in his criticisms of aspects of design.

The demise of high fluxes of gravity waves
Under these circumstances it is not obvious how the credibility of the high flux case fell so low. In fact, it was not the single uncriticized experiment that was decisive: scientists rarely mentioned this in discussion. Obviously the sheer weight of negative opinion was a factor, but given the tractability, as it were, of all the negative evidence, it did not *have* to add up so decisively. There was a way of assembling the evidence, noting the flaws in each grain, such that outright rejection of the high flux claim was not the necessary inference. After all, Weber had spent more time, effort and dedication on the work than anyone else, as some other scientists recognized. One respondent reported:

> . . . about that time [1972] Weber had visited us and he made the comment, and I think the comment was apt, that 'it's going to be a very hard time in the gravity wave business,' because he felt that he had worked for ten or twelve years to get signals, and it's so much easier to turn on an experiment and if you don't see them, you don't look to find out why you don't see them, you just publish a paper. It's important, and it just says, 'I don't see them'. So he felt that things were going to fall to a low ebb . . .

Another experimenter, who had worked with Weber, and was sympathetic to him commented:

> . . . [a major difference between Weber and the others is that Weber] spends hours and hours of time per day per week per month, living with the apparatus. When you are working with, and trying to get the most out of things you will find that, [for instance] a tube that you've selected, say one out of a hundred, only stays as a good noise tube for a month if you are lucky, but a week's more like it. Something happens, some little grain falls off the cathode and now you have a spot that's noisy, and the procedures for finding this are long and tedious. Meanwhile, your system to the outside looks just the same.
> So lots of times you can have a system running, and you think it's working fine, and it's not. One of the things that Weber gives his system, that none of the others do, is dedication — personal dedication — as an electrical engineer which most of the other guys are not . . .

Weber's an electrical engineer, and a physicist, and if it turns out that he's seeing gravity waves, and the others just missed it, that's the answer, that they weren't really dedicated experimenters ... Living with the apparatus is something that I found is really important. It's sort of like getting to know a person — you can, after a while, tell when your wife is feeling out of sorts even though she doesn't know it.

Weber himself remarked that an important factor

is having someone who is dedicated, who wants to work on the experiment until he's sure it's working properly. I think that's a key issue. I can't recall ever having set up a complex experiment which worked well when it was first turned on ... With the sort of atmosphere and the sort of situation [which we have now] people aren't likely to put themselves out to confirm the earlier results ...

Scientists' awareness of this aspect of experimental work might make them reluctant to put an overall negative construction on a set of negative results. It is clear that the possibility that the high flux claim might still be maintained in the face of this negative evidence was a significant motivating force behind the work of one critic.[4]

Crystallizing the evidence

I will refer to this critic as 'Q'. He had built one of the smallest antennae, though he argued that it was at least as sensitive as Weber's because of its sophisticated design. Nevertheless, other critics nearly always discussed the Q- experiment with reservations because it was so small. But its impact was high because of the way it was presented. As one scientist put it:

... as far as the scientific community in general is concerned, it's probably Q's publication that generally clinched the attitude. But in fact the experiment they did was trivial — it was a tiny thing ... But the thing was, the way they wrote it up ... Everybody else was awfully tentative about it ... It was all a bit hesitant ... And then Q comes along with this toy. But it's the way he writes it up you see.

Another scientist said:

Q had considerably less sensitivity so I would have thought he would have made less impact than anyone, but he talked louder than anyone and he did a very nice job of analysing his data.

And a third:

[Q's paper] was very clever because its analysis was actually very convincing to other people and that was the first time that anybody had worked out in a simple way just what the thermal noise from the bar should be ... It was done in a very clear manner and they sort of convinced everybody.

The first negative results had been reported with circumspection. Scientists discussed all the logical possibilities that could account for the discrepancy. That Weber's results were spurious was not a publishable certainty. Following closely came Q's outspoken second experimental report with its careful data analysis and the claim that its own results were 'in substantial conflict with those reported by Weber'. Then, as one respondent put it, 'that started the avalanche and after that nobody saw anything.'

The picture that emerges is that the series of experiments made strong and confident disagreement with Weber's results openly publishable but that this confidence came only after what one might call a 'critical mass' of experimental reports had been built up. This mass was 'triggered' by scientist Q.

Q believed from the beginning that Weber was mistaken, and he acted on that belief. Only the most superficial reading would lead to the conclusion that Q's actions were any less sincere than Weber's. It should also be noted that Q had prepared a strategy if high fluxes of gravity waves were found and thus was less closed-minded than a quick reading of the following would suggest. These qualifications should be borne in mind as Q's actions are analysed

These were the other important interventions by Q: the person who actually discovered Weber's computer programming mistake was prepared to allow the matter to remain private so long as Weber cleared things up quickly. However, Q forced discussion out into the open at a conference.

Their discoverer remarked of these mistakes:

> As regards the [ ] conference, Q forced my hand, I went to the [ ] conference not intending to mention the computer error unless Weber made a mis-statement . . . But when I got there Q presented me with a copy of his remarks already written up, and since I was heading off the session . . . I didn't get any lunch that day putting in to what I was going to say what happened, in what I felt was an accurate way without being emotional . . . that was the first public announcement.

Another scientist commented:

> . . . I felt that was a very inflammatory issue. It was clearly a case where Weber had tripped himself up because of his data analysis and I felt that it spoke for itself and that those few people who knew about it were enough. But 'Q' did not feel that way and he went after Weber . . . and I just stood on the sidelines covering my eyes because I'm not really interested in that kind of thing, because that's not science.

Q also wrote a 'letter' to a popular physics journal which included the paragraph:

> [it was shown] that in a . . . [certain tape] . . . nearly all the so-called 'real' coincidences . . . were created individually by this single programming error. Thus not only some phenomenon besides gravity waves *could*, but in

fact *did*, cause the zero-delay excess coincidence rate [in this data]. (Q's stress)

and that

... the Weber group has published no credible evidence at all for their claim of detection of gravitational radiation.

Q explained his experimental strategy to me as follows:

... what we could have done in the beginning was simply to have analysed Weber's performance and to have shown in principle that he couldn't have detected the gravity waves that he said he was detecting ... We could have argued from the abstract that he couldn't have been detecting them even under ideal circumstances. But we felt that we wouldn't have any credibility if we did that ... and that the only way we could get standing was to have a result of our own.

After completing work and publishing the report on their 'tiny' antenna, the Q-group built a second antenna of greater size and sensitivity but small enough to utilize the same peripheral equipment (vacuum chamber etc). I was interested in their reasons for going ahead with this, if they considered that their first antenna, though small, was large enough to do the job of legitimizing their disproof of Weber's results,

Q himself answered simply in terms of maximum utilization of available equipment. The new experiment cost next to nothing and pushed the upper bound of possible gravity waves down still further. However, another of Q's group answered:

... well we knew what was going to happen. We knew that Weber was building a bigger one and we just felt that we hadn't been convincing enough with our small antenna. We just had to get a step ahead of Weber and increase our sensitivity too.

At that point it was not doing physics any longer. It's not clear that it was ever physics, but it certainly wasn't by then. If we were looking for gravity waves we would have adopted an entirely different approach. [eg, an experiment of sufficient sensitivity to find the theoretically predicted radiation] ... there's just no point in building a detector of the [type] ... that Weber has. You're just not going to detect anything, [with such a detector — you know that both on theoretical grounds and from knowing how Weber handles his data] and so there is no point in building one, other than the fact that there's someone out there publishing results in *Physical Review Letters* ... it was pretty clear that [another named group] were never going to come out with a firm conclusion ... so we just went ahead and did it ... we knew perfectly well what was going on, and it was just a question of getting a firm enough result so that we could publish in a reputable journal, and try to end it that way.

The last phrase in the above quotation is particularly significant. Q's group had circulated a paper by Irving Langmuir (1953) to other scientists and to Weber himself. This paper was quoted to me also. The Langmuir paper deals with several cases of 'pathological science' — 'the science of things that aren't so.' Q believed that Weber's work was typical of this genre; he tried to persuade Weber and others of the similarities. Most of the cases cited by Langmuir took many years to settle. As a member of Q's group put it:

> We just wanted to see if it was possible to stop it immediately without having it drag on for twenty years.

They were worried because they knew that Weber's work was incorrect but they could see that this was not widely understood. Indeed the facts were quite the opposite. To quote again:

> Furthermore Weber was pushing very hard. He was giving an endless number of talks ... we had some graduate students — I forget which university they were from — came around to look at the apparatus ... They were of the very firm opinion that gravity waves had been detected and were an established fact of life and we just felt something had to be done to stop it ... It was just getting out of hand. If we had written an ordinary paper, that just said we had a look and we didn't find, it would have just sunk without trace.

In sum, Q and his group set out to kill Weber's findings in the shortest possible time. There is no reason to believe that they had anything but the best motives for these actions but they pursued their aim in an unusually vigorous manner. They did their experiment with the intention of developing a position from which they could more effectively destroy Weber's claims. They would probably not have bothered to carry out any experimental work if it hadn't been that they: 'looked at what some other people were planning to do and decided that there wasn't anybody who was going to make this confrontation.'

Thus, Q acted as though he did not think that the simple presentation of results with only a low key comment would be sufficient to destroy the credibility of Weber's results. In other words, he acted as one might expect a scientist to act who realized that evidence and arguments alone are insufficient to settle unambiguously the existential status of a phenomenon.

I have indicated how the experimenters' regress was resolved in this case. The growing weight of negative reports, all of which were indecisive in themselves, were crystallized, as it were, by Q. Henceforward, only experiments yielding negative results were included in the envelope of serious contributions to the debate. After Q had made his contribution to the transformation in socially acceptable opinion there simply were no high fluxes of gravity waves. Henceforward, all experiments that produced positive results, such as

Weber's, must, by that very fact, be flawed. Owning a gravity wave detector was now much more like owning a TEA-laser.

That was one way in which the field changed between 1972 and 1975. We will now look at some other changes.

Content of arguments and the nature of gravitational radiation

Scientists were ready to offer explanations for the differences in the results of the various gravity wave experiments; these ranged from estimates of relative sensitivity through to the personal qualities of the respective experimenters. The complete list of variables that were suggested to me in 1972 as candidates for explaining the differences in results was as follows:

One: *The means of detecting vibration in the bar*

As has been mentioned, Weber used piezo-electric crystals glued around the centre, but other possibilities in use at the time included sandwiching the crystals between parts of the vibrating mass, in different ways, or using a capacitor whose plate separation changed with changes in length of the bar.

Two: *The material of which the bar was constructed*

Some materials are more efficient than others. Some of the latest experiments use single huge crystals of sapphire as the vibrating mass, but in 1972 most experiments used aluminium alloy. The bars did seem to vary according to who had manufactured them and how they had been treated. One experimenter used pure aluminium, which should have been more efficient than alloy, but he had extreme problems with 'creep'; this produced a high level of 'noise'.

Three: *The electronics used to process the signals*

It was suggested that the electronic circuits could be producing the 'signals' themselves, or swamping them in their own noise, or contributing to the appearance of simultaneity in signals from different detectors, or acting as receivers for spurious non-gravitational disturbances.

Four: *The statistical techniques used to extract 'signal' from 'noise'*

As has been pointed out, decisions have to be made regarding the criterion which separates out 'signals' from noise. In the most crude systems, decisions are made by looking at a 'print-out' which will show a number of 'peaks'. Peaks above a certain predetermined level will be counted as 'signals'. This selection process may be done by panels of judges, by the experimenter himself, or by other scientists. Alternatively, computers may be used to do this analysis or an analysis using more sophisticated statistical techniques.

Arguments about the efficiency of these different techniques figured, and were to figure, largely in the debate.

Five: *The estimates made of the frequency of 'accidental' peaks and frequency of sensitive states of detector*
As has already been explained, a certain number of spurious peaks are to be expected, due entirely to noise in the system. Estimates must be made of this frequency in order to estimate the number of genuine peaks. Even where separated detectors are looking for coincident signals a certain number of coincidences will be spurious.

However, it is also the case that not all genuine gravity waves will register as coincident peaks. This is because, if the algorithm for extracting genuine peaks is based upon registering only those peaks above a certain threshold, gravity waves which pass when the noise in the bar is at a (random) low point may not excite the bar sufficiently to raise the level of energy above the threshold. Thus coincident peaks, caused by gravity waves, will occur only when both detectors are, by chance, in a sensitive state when the wave passes. Hence the number of coincidences on two detectors must be expected to be less than the number of gravity wave pulses.

Similarly, where more than two detectors are in use, the number of coincidences registered will be still less. Estimates of all these factors affect the conclusions about the number of gravity waves being registered. A high estimate of accidental signals and coincidences will leave no signals to be accounted as gravity waves.

Six: *The frequency of the radiation and the sensitive frequency of the bar*
As has been explained, the resonant bar type devices are most sensitive at their resonant frequency. Not all the experiments used the same resonant frequency so that such a difference might explain differences in results.

Seven: *The length of the bursts of radiation*
Different detectors and statistical algorithms are more or less sensitive to pulses of radiation of different length and wave form. Thus some detectors would not 'notice' very short bursts even if these bursts contained a lot of energy. This was an argument that was to figure largely later on in the debate.

Eight: *Calibration of the apparatus*
In 1972 some scientists complained that Weber had not given sufficient details of the calibration of his detector and thus that it was impossible to be certain about its sensitivity. This is an argument which grew very greatly in importance and, as will be seen below, has significance for the experimenters' regress.

Other arguments were used to try to explain how Weber's findings could be made compatible with broad cosmological considerations or why they might be spurious.

Nine: *The proximity of the source of radiation*
If the source of the radiation were near then large amounts of

gravitational energy might be detectable on Earth without implying the embarrassingly large estimates of absolute energy associated with distant sources. The intensity of gravitational radiation, it is assumed, varies over distance in accordance with the inverse square law.

Ten: *The band width of the radiation*

If the incoming radiation were all centred on a narrow band around the 1661 Hertz looked at by Weber then, again, embarrassingly high estimates of energy would be avoided.

Eleven: *Focusing of gravitational radiation*

If the assumption of an isotropic universe is dropped, and it is assumed that gravitational energy is being focused in some way toward the Earth, then again the large estimates can be avoided.

Twelve: *Spurious effects*

Scientists suggested (in 1972) that coincidences between Weber's detectors might be explained by currents in the ionosphere, neutrino fluxes, electric storms and sun spots. By 1975 broadcast television or radio waves, among other things, had been added to the list.

Some less orthodox explanations were also put forward in 1972. These included:

Thirteen: *In some way pulses of gravitational radiation are triggering the release of energy stored in the bar.*

This is a developed version of the generalized notion that gravity waves are coupled to material more strongly than had been thought.

Fourteen: *Explanation of the results of these experiments might require reference to a 'fifth force'.*

That is, some force in addition to the currently known magnetic, gravitational, strong and weak forces.

Fifteen: *The gravity wave findings are solely products of mistakes, deliberate lies, or self-deception.*

Sixteen: *The explanation might require reference to psychic forces.*

This suggestion attributes the non-accidental peaks to (for instance) the desires of an experimenter operating through psychokinesis — the power of mind over matter. A rumour circulated that Weber consulted with J. B. Rhine (the figurehead of scientific research into the paranormal), though both parties deny this. Another experimenter had seen a signal on his detector for the first time immediately after a telephone conversation with Weber and had toyed with the idea that some psychokinetic effect had been involved. Another two or three experimenters had taken an interest in research on 'extra sensory perception' and related effects. Some members of the Parapsychological Association were most interested in the work, and were delighted because they believed that some of the scientists involved were considering the psychokinesis hypothesis seriously. (This was reported to me while I was engaged on research into parapsychology — see Chapter Six. At the time I had no idea that there was any connection between the two fields; far from being a response to any question of mine, the report came as a total surprise to me.) Finally, an experiment was being planned with the collaboration of one of the secondary experimenters to test the ability of a gifted psychic on the apparatus.

None of this later group of explanations ever appeared in print — at least, not under the author's name.

By 1975, the vast majority of these candidate explanations for the discrepancies between one experiment and another had disappeared from the world of scientific discourse. The last four seemed quite bizarre and the range of discussion was restricted, as we have described, to questions of statistical error and the like.

This is exactly the sort of change we would expect to take place as the field reached consensus. As the disturbance brought about in the scientific community by the initial claims was smoothed away, there was no further need to dig deep into the background of 'cherished beliefs' to try to bring a new order to physical reality. Weber was simply wrong. The still, deep waters of everyday life lapped back over the volcano that had thrust through their surface. Today its place is marked by scarcely a ripple.

Consider if the argument had gone another way. Suppose, for a moment, that it was devices that detected gravity waves that had come to be defined as the competent designs. In that case high fluxes of gravitational radiation, or whatever it was that was causing the coincidences in Weber's detectors, would have been *defined* as something that could be seen on an apparatus like Weber's but not seen on the apparatus of his critics. The differences between the two sets of antennae — those that could detect the phenomenon and those that could not — would now *explain* the nature of whatever it was that was causing the coincidences. That is to say, whatever was causing the coincidences must be something of a nature that could affect Weber's antenna but could not affect the antennae of his critics.

For example, take point six above: if the critics were working on a different frequency range, then we would know something about the frequency distribution of the radiation: it must be restricted to the Weber waveband.

If point seven were the crucial difference then they would provide clues about the pulse shape of the radiation, a point which Weber tried to establish.

If the different performances were explained in some way by the catalogue of possibilites under point twelve, then some new, non-gravitational, phenomenon might be what had been discovered by Weber.

Finally, discovering that only points thirteen or sixteen could account for the difference in performance between the two sets of antennae would entail something like a revolution in physics. (To talk about 'discovering' such things is disingenuous; one should rather talk of establishing or 'negotiating'. In Chapter Six these different negotiating strategies will be discussed at length.)

As I have explained, the radical possibilities that Weber's work suggested had disappeared from the collective consciousness of physicists by 1975. By then the nature of Weber's claims had been settled: they were simply a mistake of no significance. In the counter-factual circumstances something more startling might have been revealed. I am arguing here that just as the process of deciding whether gravity waves had been detected was coextensive with deciding which set of results was to be believed, so the detailed *nature* of gravity waves was settled at the same time. Different decisions about the quality of the experiments would have gone hand-in-hand with different decisions about the nature of gravity waves. This can be summed up as a ninth proposition:

> *Proposition Nine:* Decisions about the existence of phenomena are coextensive with the 'discovery' of their properties.[6]

An attempt to break the regress:
the calibration of experiments
Though the demise of gravity waves has been largely explained, it is worth examining other attempts that were made to break the experimenters' regress. Various non-experimental and 'non-scientific' activities can be seen in this way; if viewed in this way, the conspiratorial activities of some of the scientists trying to discredit high fluxes of gravity waves by discrediting Joseph Weber himself seem much less surprising. The reader should now turn back to the first, third and fifth set of comments of gravity wave researchers in 1972 and look at them this way; it is a matter of reaching for anything in the absence of independent criteria. These 'unscientific' solutions serve a similar purpose to the most abstract theoretical arguments about general relativity or the nature of the cosmos. Both conspiracy and *a priori* theories are attempts to break the regress.[7]

Another episode of the gravity wave story not only illustrates Proposition Nine but shows the circular nature of the regress very clearly. This was scientists' attempts to institute a 'test of a test' as a way of settling the argument. This test was to be the calibration of the competing antennae. If it could be shown that Weber's relative sensitivity was not as great as he claimed, then negative experimental results would have more credibility.

The calibration of instruments is a familiar procedure. Imagine that a prototype voltmeter has been constructed. It consists of a needle which swings across a scale but as yet the scale is blank. To calibrate the instrument known voltages are applied to the terminals and the corresponding positions at which the needle comes to rest are recorded. Thus, marks corresponding to known voltages can be inscribed on the scale. Henceforward the meter may be used to measure unknown voltages; the unknown voltage is applied to the terminals and the mark against which the needle comes to rest gives the answer.

The assumption built into this procedure is that the unknown voltage acts upon the meter in the same way as the standard voltages which were applied to calibrate it. This is so slight an assumption as hardly to be worthy of the name. After all, a voltage is a voltage is a voltage! Nevertheless, it would be correct to say that during the calibration of a voltmeter, standardized voltages are used as a surrogate for as yet unmeasured signals. In more controversial science the assumptions underlying the process of calibration are of greater moment.

Calibrating gravity waves

Some of Weber's critics, in an attempt to short circuit arguments about the relative sensitivity of different experiments, physically calibrated their antennae. They did this by injecting pulses of energy into the bar via an electrostatically energized 'end-plate'. The end-plate could inject tiny vibrations into the bar in a well-understood way. What this calibration procedure amounted to was the use of the antennae to detect a well-understood surrogate phenomenon. It was clear to all that what counted as a well-designed instrument, as defined by this test, was one that would detect the electrostatic pulses; there was no question as to these pulses' existence.

Weber was initially unwilling to calibrate his own antenna in this way. A critic of Weber described the situation as follows:

> We had calibrated our own antennae in a unique way which depended in no way on calculation. So we knew what our sensitivity was and at that time we could only calculate what Weber's sensitivity was. So you're right in saying that the relative sensitivity was something that was on the one hand calculated and on the other hand known to absolute accuracy ... Soon thereafter we did get an opportunity to go down and calibrate Weber's antenna and we found ... our calculations were correct...

As this respondent suggests, the outcome of the tardy electrostatic calibration of Weber's apparatus was seen by most to be a vindication of the critics' calculations. It was felt to be a decisive demonstration that the sensitivity of the critics' antennae was at least as great as Weber's. In particular, an argument concerning the correct way to process incoming signals seemed to be settled. Weber insisted that the maximum net sensitivity was to be obtained by a non-linear, or energy, algorithm (the algorithm relates to the circuitry and computer program which processes the raw signal). Weber's critics insisted that a linear, or amplitude, algorithm was the best and uniformly made use of the linear algorithm. As one respondent put it:

> For a signal with a sine wave underlying . . . it turns out that a system which is linear can be shown theoretically and quite soundly to be the best system for detecting things. But Weber's always used the non-linear system and so his initial claim was that it was just clearly superior because he finds gravity waves with it whereas people with the linear systems don't. Despite the fact that you can prove rigorously that it's not so.
>
> Well Weber was pushed very hard on this and he finally implemented both systems . . . and he hooked up to the same detector both a linear system and a non-linear system . . . and what he found was that he did indeed find gravity waves more often with his system. However finally after much pushing he put calibrators on — things that could simulate gravity waves — and it turned out that the linear system was about twenty times better at finding the calibrator signal. . . .

In this quotation it is the last phrase that is the most important. Weber did not accept his critics' interpretation of the calibration results. Instead, he claimed that the form of the calibration was inappropriate. Thus:

> *Collins:* In reading your 1974 publication I understand that you did a calibration experiment using both algorithms and that you got a better result with the linear algorithm?
> *Weber:* No that's not right. The linear algorithm used by other people is unquestionably superior for short pulses — let me make that absolutely clear. There are certain arguments given for use of the linear algorithm. These arguments are applicable to short pulses and in my opinion they are correct arguments. And the fact that the linear algorithm is not in fact more sensitive is giving us information about the character of the pulse. It means that the character of the pulses do not fit the assumptions which went into that method of analysis . . . so far we think of several kinds of signals that would give results somewhat similar to the ones we see.

Weber's critics read this claim in a less positive light: one remarked:

> What he did was to change the nature of the signals. He said: 'Well, the signals must not be of the form which we've been assuming. They've got to be something else now.' Some strange waveform of which he failed to give a single example. 'And so my algorithm is now best again.' In fact that resolved a lot of difficulties for him. He was wondering why we didn't see his signals. And he said, 'Now I know why. The signals are of a weird form.'

Another respondent, remarking on the failure of Weber's algorithm in the calibrator test, said:

... you have this incredible conflict that when you look for gravity waves the other system seems to do a better job — that's a perfect example of a negative experiment done by the author. It demonstrates that there's nothing there.

One might describe these arguments as turning on the appropriateness of the surrogate signal used for calibration purposes; the assumption that is hardly worth calling an assumption when calibration is carried out in 'normal science' takes on considerable salience in this case.

The force of assumption

To go straight to the end of the story, Weber's interpretation of the calibration results were greeted with scepticism. Weber did manage to invent hypothetical signals compatible with the calibration test; they had a pulse profile such that they would be more easily detected by his antenna, using his algorithm, than by his critics' methods. However, the existence of such signals was thought unlikely by most scientists. According to one respondent, signals with such a profile were 'pathological and uninteresting'. In other words, it would be difficult to think of cosmological scenarios that would give rise to signals with such strange and exact signatures. In the current state of cosmology Weber's hypothesized signal shapes were too 'implausible' to be considered seriously.

To sum up, because of the implausibility of Weber's account of the reasons for the unsuitability of the surrogate calibration signal, the calibration episode made a contribution to the closure of the debate and helped to bring about the demise of high fluxes of gravity waves.[8]

There is, however, a little more that can be said about the case. It was not only the failure of Weber's *ad hoc* hypotheses that allowed the closure, but also the very act of calibration itself. In retrospect, Weber would have served his case better to have maintained his refusal to use electrostatic calibration — not just because the results proved unfavourable but because of the assumptions taken on board by the act of calibration and the restrictions of interpretation imposed as a result.

In bowing to the pressure to calibrate electrostatically, Weber set at least two assumptions beyond question. First he accepted that gravitational radiation would interact with the substance of his antenna in the same way as electrostatic forces. This is certainly a slight assumption, yet, as has been shown in this chapter, there were times when informal discussion took place as to whether gravitational force might be coupling more effectively than expected with the matter of the bar through the release of latent energy via a mysterious mechanism.

More importantly, Weber put it beyond question, at least for the time being, that the insertion of a localized pulse into one end of a bar antenna would have a similar effect to the insertion of energy into the bar as a whole from a source at a great distance. Again, this might seem a slight assumption — clearly it was one that Weber did not feel able to dispute — yet more recent events show that it is not inviolable.

An alternative surrogate

An experimenter working on a more modern antenna, whom I interviewed in 1980, planned a different type of calibration. He intended to use as a surrogate not electrostatic force but the fluctuating gravitational attraction induced by a small spinning bar of material located close to the antenna. The rapid changes in gravitational attraction between the material of the antenna and the material of the bar, as their relative dispositions changed, was intended to mimic the high frequency changes in the gravitational attraction of objects for one another symptomatic of gravitational radiation. Though this respondent's apparatus was of a more complex, non-resonant, design than Weber's, his answers to questions about calibration methods are germane:

> *Collins:* What is the advantage of the spinning bar calibration over electrostatic calibration?
> *Respondent:* Well since it couples gravitationally to the antenna it does give you a somewhat more basic measurement — if you like — it's still not really what you want. It still doesn't duplicate the effect of gravitational radiation because it's a near-field effect and the spinning bar really only couples to one end of the thing instead of coupling uniformly to the entire antenna. So this is the limitation of this sort of approach. The spinning bar is more appropriate with something like a Weber resonant antenna where you can more nearly couple to the antenna...
> *Collins:* How certain can you be that electrostatic calibration pulses are acting as an exact analogue of gravity?
> *Respondent:* Oh, they're not. They're certainly not ... From simple measurement [using electrostatic calibration] ... I know precisely the force that I'm applying ... and I can calculate the size of the signal that I ought to get out of the transducers and that's all. But it doesn't mimic the effect of a gravitational wave on the antenna. And that's true whether it's this sort of antenna or whether it's a resonant bar. The fact is that the gravitational wave interacts with all parts of the antenna, with all of the mass of the thing, and there's just no way of reproducing that — at least there's no way I'm able to think up of producing that effect...
> What you are trying to do with electrostatic calibration is to check your theoretical calculations ... What you can't test in this way is the theoretical calculation that tells you precisely what happens when a gravitational wave of a certain amplitude hits the antenna...

For this respondent, with his more complex antenna and his idea for a different method of calibration, the assumptions underlying electrostatic calibration were worth analysing and circumventing if possible. He had thought of a way of circumventing the need for electrostatic impulses using, instead, changes in the gravitational attraction of a local mass. He was still unhappy with the need to use a localized source rather than a powerful distant source that would more nearly mimic effects of gravitational radiation on his antenna. Though he felt that electrostatic calibration would not be as poor a substitute in the case of a Weber bar as it was for his own apparatus, this was only because the analysis relating localized forces to distributed forces seemed simple and plausible in the case of the Weber bar. As he put it, 'there's no argument about it.'

That there was no argument is literally true as I have pointed out above. Weber in accepting electrostatic calibration chose not to argue on these fronts. My respondent's decision to open up the range of possibilities for calibration signals reveals that such an argument might not have been entirely implausible.

Calibration is the use of a surrogate signal to standardize an instrument. The use of calibration depends on the assumption of near identity of effect between the surrogate signal and the unknown signal that is to be measured (detected) with the instrument. Usually this assumption is too trivial to be noticed. In controversial cases, where calibration is used to determine relative sensitivities of competing instruments, the assumption may be brought into question. Calibration can only be performed provided this assumption is not questioned too deeply. In fact, the questioning is constrained only by what seems plausible within the state of the art of the science in question. But the very act of using a calibration surrogate may help to establish the limits of plausibility.

Weber, in accepting the scientific legitimacy of electrostatic calibration for his gravitational antennae, thus accepted constraints on his freedom to interpret results. The act of electrostatic calibration ensured that it was henceforward implausible to treat gravitational forces in an exotic way. They were to be understood as belonging to the class of phenomena which behaved in broadly the same way as the well-understood electrostatic forces. After calibration, freedom of interpretation was limited to pulse profile rather than the quality or nature of the signals.

The anomalous outcome of Weber's experiments could have led toward a variety of heterodox interpretations with widespread consequences for physics. They could have led to a schism in the scientific community or even a discontinuity in the progress of the science. Making Weber calibrate his apparatus with electrostatic pulses was one way in which his critics ensured that gravitational radiation remained a force that could be understood within the ambit of physics as we know it. They ensured physics' continuity — the maintenance of the links between past and future. Calibration is not simply a technical procedure for closing debate by providing an external criterion of competence. In so far as it does work in this way, it does so by controlling interpretative freedom. It is the control on interpretation which breaks the circle of the experimenters' regress, not the 'test of a test' itself.

Notes

1. Critics of these studies do not seem to have faced up squarely to the experimenters' regress. Some (such as Mulkay, Potter and Yearley, 1983) have not been able to recognize the wood of the argument among the trees of the data. Some (such as Laudan, 1982) have mostly complained about the unpalatability of the consequences.

2. Interestingly, the level was 2.6 standard deviations. This would count as a high level in the social sciences, but appears to have been counted as inadequate in this area of physics. The reader may recall, from Chapter Two, that choosing an appropriate level of statistical significance called for the invocation of a mysterious murine rule.

3. Production of a 'correct' result is only a necessary, not a sufficient condition for the ascription of competence. Scientists feel free to criticize their colleagues even when their results accord with their own dispositions so long as there are still other negative results to which to refer. A similar phenomenon will be reported in Chapter Five and discussed in Chapter Six.

4. A completely disinterested, honourable and impersonal motive, let me hasten to add, since this critic was firmly convinced that Weber was wrong. His reasons were drawn from physical theory and his exceptionally wide experience as an experimenter in difficult areas.

5. In a talk given to the AAAS in 1978, Kip Thorne referred to this sort of consideration as 'physicists' cherished beliefs'. These are not arbitrary beliefs of course; they are cherished only because giving any of them up would involve giving up so much more of what has proved successful in the physicists' network of concepts. This will be discussed at greater length in Chapter Six.

6. An important consequence of Proposition Nine is that the success of one party to a dispute of this sort *cannot* be explained by their superior grasp of the nature of the phenomenon under investigation. It is this that is being discovered (determined) by the debate itself (cf, Farley and Geison, 1974; Roll-Hansen, 1984).

7. Rosenthal's 'replication accounting' system (Chapter Two) can be thought of in the same way. In spite of his declared intentions of merely adding results irrespective of their origin, Rosenthal had to fall back on measures of experimental quality. For example he did calculations that rested especially heavily on supervised doctoral dissertation work and on studies that contained special controls for minimizing cheating and error. Irrespective of the quality of these categories of experiment (student work is not normally thought of as being among the best), the point is that simply to aggregate experiments touches not at all on the regress; it simply ignores the question of quality and is not a satisfactory solution.

8. One respondent's remarks show that it is only a matter of implausibility, not techical impossibility. He said:

> . . . there is one logical possibility, in a sense, and that is that gravity waves don't behave anything like we think they behave, the whole theory is complete hogwash,

and that they have some screwball properties, and by some fantastic chance Weber had just happened to build a detector that somehow or other, in some mysterious fashion, picks these up....

Technical appendix

To detect gravitational radiation, a signal must be separated from noise. Most antennae recorded their data in the form of a spiky chart recorder line. The following appendix explains the techniques for extracting the signal and some of the developments which led sceptical scientists to treat Weber's claims seriously.

Techniques and Innovations in
the Search for (hf) Gravity Waves

Figure 1   **Figure 2**
Signals as Peaks above Threshold   **Signals as Sudden Changes in Energy**

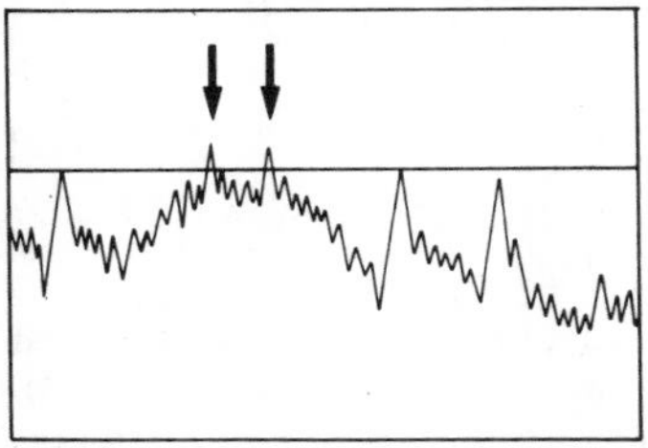

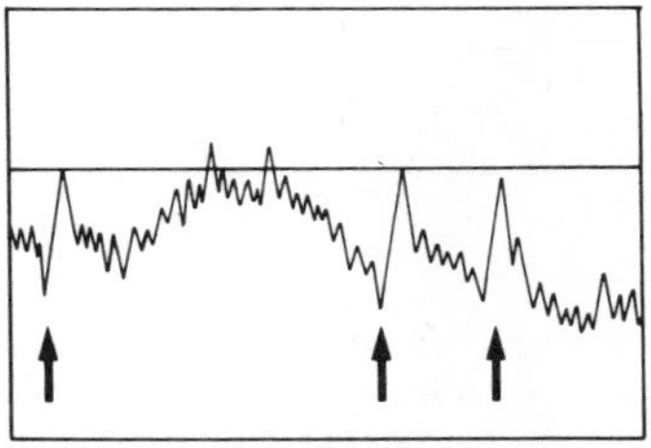

Even the most well isolated detector will produce a 'noisy' output because of thermal noise in the aluminium alloy bar. Some method of extracting signal from noise has to be used. In the early days Weber counted each peak above a predetermined threshold as signifying a wave pulse (Figure 1). An alternative is to look for sudden changes in the energy of the bar, irrespective of whether or not a threshold is crossed (Figure 2). The latter seems to be a more efficient method. Weber's early analyses of his output were done by 'eyeball'. This was a widely distrusted aspect of his design, though it can be defended. (After all, the eye is much better at pattern recognition than is a computer.) All the later experiments used a computer to do a 'hands off' analysis of data.

Figure 3
Signals as Coincidences from Two Detectors

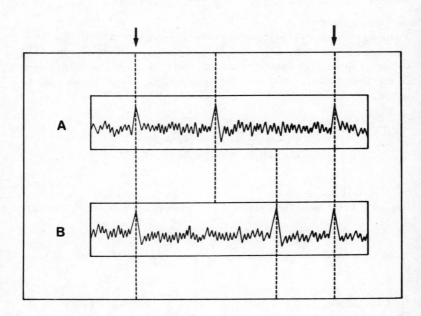

A major innovation was the comparison of output from two (or more) isolated antennae. Antennae A and B produce output traces that are compared (Figure 3). Only coincident peaks (arrowed) count as genuine gravity waves. There is still the problem that a few coincident peaks will occur because of coincident peaks of noise in the two detectors. These are known as 'accidentals'. Accidentals and genuine peaks can be separated with a 'delay histogram' analysis.

Figure 4
Signal Extracted from Noise by Delay Histogram

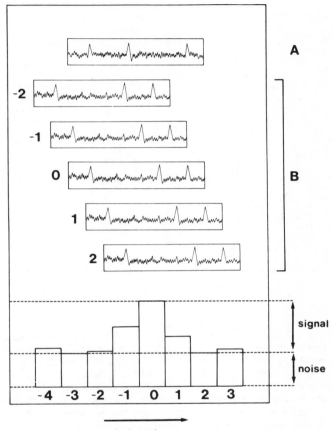

relative time

The delay histogram (Figure 4) is constructed by taking the output from antenna A and comparing it with output from B when that output is displaced in time by varying amounts. When the time displacement is large, the coincident peaks on the outputs should be the product of noise alone. An estimate of the number of accidentals can thus be obtained from the height of those histogram bins which are far from the centre (zero time displacement) of the delay histogram. The signal is then represented by the height of the central bin less the height of the background accidentals. Since the time resolution of bar antennae is not perfect, signals will spread slightly in time, so that bins near to the centre of the delay histogram should register above the noise level.

Figure 5
Periodicity in Signal

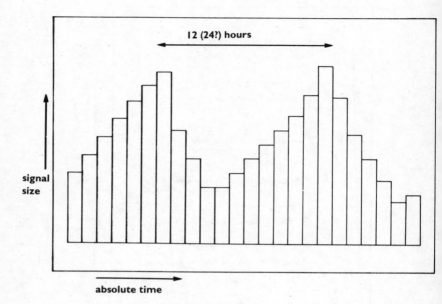

If the excess signal over noise is determined for each hour of the day and night, and the hourly totals are aggregated for each hour over a period of weeks or months, a periodicity may be noticed. The histogram in Figure 5 shows the results of such an exercise and reveals a periodicity with about a 12 hour cycle. In the early days, Weber claimed to find a periodicity with about a 24 hour cycle. He reasoned that if gravity waves come from one point in space (for example, a point where there are a lot of stars — such as the centre of the Galaxy) then, as the Earth rotates, an antenna fixed to its surface will be in a disposition that is most efficient for detecting radiation from that direction once per Earth rotation — that is, about once every 24 hours. It was then pointed out that since the Earth is virtually transparent to gravitational radiation the efficient disposition would be attained twice in each rotation (once on each side). Later Weber claimed that the periodicity had, in fact, about a 12 hour cycle.

Figure 6
The Sidereal Correlation

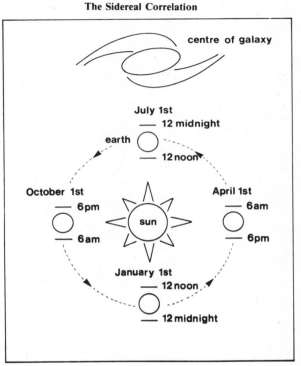

But if the centre of the Galaxy is the source of the radiation, rather than the Sun, the phase of the periodicity should change during the year (Figure 6). (In other words, the astronomical day is nearer 23 hours and 56 minutes.) Thus if the disposition of the antenna (represented by the same straight line on the surface of the Earth) is most efficient at 12 noon and 12 midnight on 1 January, it should be most efficient at 6.00 am and 6.00 pm on 1 April, noon and midnight again on 1 July, and 6 and 6 again on 1 October. This phase shift is the 'sidereal correlation'.

Chapter Five

Some Experiments in the Paranormal: The Experimenters' Regress Revisited

Gravitational radiation research is a branch of physics. Its theoretical bases are modern cosmology, the general theory of relativity, and those branches of science and mathematics having to do with the design of the detector itself. The work described in the last chapter may have been on the frontiers of science, but they were the frontiers of mainstream physics. Many of those who participated were highly placed physicists in prestigious research institutions. One or two very well known scientists either did experiments or contributed to the theoretical debate.

Weber's heterodox findings opened up the conventional world of physics for a short time, but the skin soon healed over the wound. The next case study, by contrast, is in an area very much removed from the centre of scientific research. Indeed, it is about as far from the centre of scientific research as it is possible to get within the current usage of the term 'science'. Some might say it was a little beyond the frontier. The subject is the emotional life of plants.

Responses of plants to remote stimuli

Toward the end of the 1960s Cleve Backster, a lie detector expert in New York, attracted considerable publicity for discovering, as the newspapers put it, that plants have feelings. Lie detectors work by registering changes in the electrical conductivity of the human skin and are widely used in the United States. Backster was a sufficiently well established expert to be asked to present a statement to a congressional hearing on the subject in June 1974.

The story of his less orthodox work began, as he reports, when, in an idle moment, he attached the electrodes of a lie detector to the leaf of a philodendron plant and was surprised to discover that sudden changes in the resistivity of the leaf were registered. These changes seemed to be correlated with his mere *intention* to damage the plant, by burning it and so forth, even though he had not yet damaged it physically. Subsequently, Backster decided to test these findings in a controlled manner. The first formal presentation of his results can be found in *International Journal of Parapsychology*, entitled 'Evidence of a Primary Perception in Plant Life' (Backster, 1968). Fairly full but uncritical accounts of this work may be found in the first three chapters of *The Secret Life of Plants* (Tomkins and Bird, 1974).

This type of experiment can be done by measuring either of two different electrical characteristics of the plant. In all the early experiments the resistance of the plant material was measured using a lie detector (polygraph). A lie detector consists essentially of a sensitive Wheatstone Bridge (resistance measuring) circuit, an electrode system to link the circuit to the plant, and some means such as a chart recorder for recording changes in resistance. This version of the experiment involves passing small currents through or across the surface of the plant leaf. Either direct or alternating current may be used to measure resistance in this way. An alternative design — used in some later experiments — involves an electroencephalograph-type circuit (EEG) which measures the electrical output of the plant rather than its resistivity. This design does not require that a current be passed through the leaf.

In 1968 Backster reported that he had arranged an automatic apparatus to drop a series of live brine shrimps into hot water, thus killing them. During a sequence of drops, chart recorders monitored the electrical resistance of the leaves of three living plants. Backster claimed that a statistically significant correlation was found between brine shrimp 'termination' and chart recorder activity; no such relation was to be found when simulated 'drops' were made without live shrimps. The live drops and simulated drops were alternated in a random sequence. Backster's reports eventually attracted enough notice to cause other scientists to criticize and later to repeat his experiments.

Fieldwork

My investigation of this area started in 1972 and included interviews with Backster and two other scientists who had claimed to find positive results in similar experiments. I carried out a second field trip in 1975.

Before setting out in 1972, I 'repeated' Backster's experiment myself. With help from colleagues in the physics department of my university, I attempted to measure changes in the resistance of the leaves of a plant. We made up a pair of small aluminium plates that could be clamped either side of a leaf and connected these to a Wheatstone Bridge circuit. We monitored the measured resistance of a leaf attached to a living plant and recorded the changing output on a chart recorder. Then we administered emotional stimuli to the plant. These included verbal threats, bringing a lighted match close, and so forth. Our chart recorder showed large fluctuations in leaf resistance but none of these could be correlated with any of the stimuli. Our conclusions were pessimistic.

The 1972 field trip

As well as positive experiments, one negative experiment had been reported by 1972. A short account of this can be found in *The Journal of Parapsychology* (Johnson, 1972) in the form of a letter from R. V. Johnson. Johnson wired up his plant in the way described by Backster. He discovered similar fluctuations in leaf resistance but, just as in our own experiment, he did not find any correlation with emotional stimuli. However, Johnson reported that when he put his plant in a controlled environment the unexplained fluctuations ceased. They could be reproduced by varying the air temperature by one or two degrees centigrade or by varying the humidity by ten to fifteen percent.

Johnson claimed competence for his work at the outset by citing the authority of orthodox scientists. He wrote:

> I did my research under the careful scrutiny of the electrical engineering and botany departments at the University of Washington and received my Master's degree based on my work.

All three of the scientists I spoke to knew the Johnson experiment quite well. I asked them how they coped with what seemed to be watertight and competent criticism of their results. It turned out that they did not see the experiment in this way at all. Indeed, they were surprised that the Johnson work had received the attention it had. The following is an assorted list of replies from the three.

First, comments from two of the three respondents indicate, in a general way, their belief that Johnson's experiment had not approached the necessary levels of expertise:

> ... Most people when they don't follow the instructions exactly have the sense to keep quiet about it ...
> he just junked the experiment. In fact, not only did he junk it through the instrumentation, but it was the whole format he threw out ... So he wasn't even doing the same thing. What he did is discard my experiment as unmanageable in his preliminary observations. Which doesn't give him a voice to say anything. In other words, it's like he didn't do anything, that's all as far as science is concerned.
> Well, I read his article, and I said to myself — well this is a very fine piece of technical workmanship here but the man is not pursuing it from the proper perspective and therefore he doesn't get results and that was it. I said 'forget it' ... to me it was just another master's thesis.

These next comments are respondents' detailed criticisms of Johnson's techniques. (Note that these are the remarks of three different respondents, so they are not necessarily in agreement.)

In changing from AC to a DC bridge he lost the phenomenon completely.

... Johnson's results were quite correct but the difference is that in my experiments I'm interested in very short-term responses, for instance responses that take place in less than five seconds, so that the long-term responses reported by Johnson do not really count.

Johnson's results were nonsensical because he used aluminium foil [electrodes] and you must use one of the noble metals for contacts in order to avoid the oxidation effect, which will create artifacts from now until the aluminium foil completely corrodes away ... the effects that Johnson was measuring, the effect of humidity, CO_2 concentration, and so on, were so large as to mask the effects of the oxidation of the aluminium foil so that he got good results for what he was after. But this is nothing like sensitive enough to find the Backster effect. It's no surprise he didn't find it.

He didn't isolate the plants for a period before the experiment.

... it's completely haywire, and in fact what you must first do is make some sort of energetic connection with the plant and then you can hold the polygraph completely steady and it is no longer affected by temperature changes — that's after you've made a sort of psychical connection with the plant. Developing a psychic link between yourself and the plant is like suddenly dropping a Faraday cage round it ...

The Johnson experiment was, then, interpreted differently by different observers. This should no longer surprise the reader. Though in 1972 it was seen by the many critics of the Backster findings as a definitive refutation — it was good enough to merit a report in the *Journal of Parapsychology* — believers in the effect were able to dismiss it out of hand on the grounds of incompetence or technical deficiency.

My own experiment, which had seemed quite good at the time, now seemed totally inadequate. We had certainly not isolated the plant for a period. We had not 'tried to form a psychic connection with it'. And even if these somewhat paranormal demands were not taken seriously, we had not used silver, gold or platinum for the electrodes; and this turned out to be accepted as a vital precaution even in the respectable reaches of plant physiology. We had done no more than a failed preliminary run! We could only have come to our firm negative conclusions because our certainty that we had the recipe for the correct conduct of the experiment overruled our knowledge of what doing an experiment really involved. This accords with Proposition Six.

The 1975 field trip

Between 1972 and 1975 the results of two other negative experiments were widely publicized (eg, see Chedd, 1975). My sources in 1975 comprised Backster and these two new critics — Drs Gasteiger and Kmetz — discussions with other interested parapsychologists, various documents (including Horowitz *et al*, 1978) and a tape recording of a symposium held at the American Association for the Advancement of Science (AAAS) annual meeting in January 1975 where, among others, Backster, Gasteiger and Kmetz presented papers and answered questions.

A striking difference between 1972 and 1975 was that the Johnson work had become invisible. It seemed that now *critics and believers alike* agreed that Johnson's experiments were not sufficiently technically competent to count as refutations of Backster's results. This may have been a consequence of the general increase of technical knowledge in the field. Advances in what counts as 'expertise' certainly did take place in the intervening years, for both new negative experiments were conducted in an atmosphere of continuing consultation with Backster himself. Equipment was loaned to Kmetz by Backster and Gasteiger's graduate students met Backster to discuss the experiment.

There is no reason to suppose that either group set out with specifically negative intentions; rather they appear to have taken the experiment very seriously. Indeed Gasteiger, when pressed, was prepared to concede that the residual differences between his own and Backster's experiments could conceivably account for the differences in their results. But, as he put it, there is a point at which one can no longer continue to make further adjustments to one's experiments but must publish what has been found to date, which in this case were negative results.

Thus both these experimenters had at least tried to develop their expertise to a higher level than Johnson's. I believe they succeeded. This seems clear when their endeavours are compared to Johnson's own statements that his initial intentions had been to 'try to obtain Backster's results independent of his recommendations' (Johnson, 1971, p 3) and that

> After completion of the experiments silver foil was recommended as a better electrode material than aluminium. No experiments were attempted with silver foil though, as it was not considered important enough to change the results (Johnson, 1971, p 7)

One can only wonder at what kind of abstract ideal of replication informed Johnson and his supervisors. In contrast, both Gasteiger and Kmetz took great care over their electrodes.

The role of negative experiments

By 1975, then, Johnson's experiment was manifestly inadequate as far as all those 'in the know' were concerned. However, the very high visibility of his work in 1972 still needs explanation. A functional explanation seems reasonable. My argument here is slightly different to that of the previous chapter, where I suggested that Q had been responsible for crystallizing negative opinion against Weber. I said that Q had only done his experiment in order to give legitimacy to his theoretical arguments. I am suggesting here that the cards were so heavily stacked against Backster in the first place — his work on the vegetable kingdom was felt to be beyond the pale even by parapsychologists — that no crystallizing effort was required.[1] Johnson's experiment was used, however, to legitimate public dissent. It seems that any empirical result, however slight, can be used for this purpose if the circumstances are right.

By 1975, with the new 'definitive' refutations performed by experts, critics no longer 'needed' the Johnson result. They could afford to admit to its technical limitations. The AAAS symposium presentations had removed Backster's results from serious consideration as far as critics and the wider community of scientists were concerned.

Neither of the later negative experiments had, apparently, affected Backster's confidence in his results. This was made clear in my interview with him and in his short response to the presentation of the critical experiments at the AAAS. The latter began as follows:

> ... I feel as though it's after the crucifixion, but let's make believe it's Easter, and let's see if I can rise now ... My understanding of scientific methodology is that before you can actually fail to repeat an experiment, you must attempt to repeat it. And when you attempt to repeat it, you must precisely follow the direction of the original experimenter.... [but] I have heard time and time again of the adjustments they made because they thought it was better or easier. This is not replication in any sense of the word ... There has been no successful failure to replicate that experiment ...

Backster then listed the technical deficiencies of the negative experiments, as follows (my paraphrase):

> One experiment was performed by graduate students and as they work to deadlines, and therefore under pressure, it is probable that they were not psychologically attuned to obtaining results in the area of consciousness research.
>
> The experiments were carried out during the day, in the normal environs of a biological laboratory, and as it seems that stimuli might be 'received' from any part of the building in which the plants were kept, and as there was likely to be a great deal of biological 'noise' in the building, the plants might be expected to show no outstanding reaction to the experimental stimuli.
>
> More brine shrimp 'drops' were used in each run of the reported experiments than in the original experiment. As the plants show an

'adaptation effect' this 'dilutes' the data unnecessarily, perhaps to a statistically insignificant level.

Switching from a DC to an AC bridge might destroy the effect. '...we don't even know what this phenomenon is. We haven't even identified it. If there is a remote effect — if there is an attunement to consciousness ... how can we arbitrarily jump from one technique to another?'

Relationships between experimenter and plants were not carefully controlled in the appropriate manner before the experiments.

The monitoring equipment was used on automatic recentering mode which, it has been stressed, disturbs this experiment for unknown reasons.

It seems both from the tone of the conference, and from his own comments and others' reports, that Backster was not finally crucified *at* the AAAS symposium. Its long term effects were more significant.

The world we have lost

In case the suspicion lurks that this is a 'special case' in that Backster was being unusually, or even uniquely, perverse, unreasonable or irrational, I present the following anecdote. This involved some fast footwork on the part of Backster's critics and it may help to make available to the reader the sense of symmetry and the inconclusiveness of the debate as it took place. It is hard to recapture.

An essential part of the experiment is the control sequence, in which the automatic shrimp-dumping apparatus functions exactly as in live 'drops' except that the shrimps are absent. At the AAAS symposium a member of the audience criticized Gasteiger's experiment because tap water was dumped into the killing tank in control drops. The claim was made that, as tap water may very well contain living organisms, the plants may have been stimulated by the control drops in the same way as in the live drops. This would have accounted for Gasteiger's lack of results on comparing live with control runs. In the discussion which followed, Gasteiger defended the use of tap water. Later, in my discussion with John Kmetz, I raised this point and he too argued that this particular objection was not significant.

I have no interest in discussing the 'actual significance' of this point; however, in the account of Gasteiger's experiment which was subsequently published in *Science*, it is stated that the control drops dumped *distilled* water into the killing tank. Kmetz, in his defence of Gasteiger's tap water technique, was unaware of this later adjustment.

My discovery of this discrepancy occurred after my interview with Dr Gasteiger so I was unable to discuss it with him; the version of the source of discrepancy which is known to me is due entirely to Backster. He told me that, as far as he *now* understood Gasteiger's experiments, he used distilled water in the controls. He suggested that Gasteiger had forgotten this experimental detail at the time of the

AAAS meeting since the experiments were conducted, for the most part, by his graduate students.

Thus, as Gasteiger's conduct at the AAAS exemplifies, critics of an unpopular finding sometimes make firm defences of even imaginary details of experimental protocol where their own experimental design expresses the validity of their opponent's viewpoint. It would then be quite incorrect to characterize the debate as consisting of 'sound definitive procedures' on one side and *ad hoc* rescue acts on the other.

It is interesting that Backster, though he knew about the discrepancy in the report in *Science*, never, to my knowledge, tried to use the issue to cast doubt on the veracity of Gasteiger.

Random number generator experiments
So far all the examples, arguments and counter-arguments over experimental competence have taken the same form. In every case scientists have argued that others' experiments were deficient in one or more respects. For completeness I now include a case where the scientists involved discounted their own experiment when it achieved a negative result on the grounds of their own incompetence, while a disbeliever in the phenomenon claimed that their negative result actually proved that they had done the experiment competently!

In 1969 *New Scientist*, a British science magazine, published an account of the experiment to be discussed below. In an editorial it presented the experiment as an answer to one of parapsychology's most trenchant modern critics, C.E.M. Hansel.

> Not long ago in *ESP: A Scientific Evaluation* (MacGibbon and Kee, 1966, p 241), Professor C.E.M. Hansel concluded a critical survey of parapsychology with the words: 'If 12 months' work on VERITAC (a machine used by the US Airforce Research Laboratories) can establish the existence of ESP the past research will not have been in vain ...' Dr Schmidt's machine seems in no way inferior to VERITAC and he has now come up with positive results. It remains only for other investigators to confirm these findings. (16 October 1969, p 107)

Dr Helmut Schmidt, a physicist working for Boeing, had reported the construction of a random number generator machine. Four circuits, linked to a four lamp display, were opened or closed at random as a function of the decay of a small quantity of radioactive material (Schmidt, 1969a and b). Subjects could complete the circuits to the lamps by pressing one of four buttons placed beneath them. One of the four lamps would then light, the particular one being determined by the position of the radioactive decay-controlled switch. Subjects would aim to 'guess' the position of this switch by pressing the button beneath the lamp they had predicted would be the next one to light. They should succeed one time in four on the basis of chance expectation.

Details of the device itself, and details of tests for randomness of the output when the buttons were pressed automatically, were published in an article in the *Journal of Applied Physics* (Schmidt, 1970), a prestigious orthodox physics journal. *New Scientist*, and other publications gave details of the performance of experimental subjects who had attempted to press the appropriate buttons. (Hits and misses were recorded automatically.) During 20,000 trials four subjects among them scored at about eight percent away from chance expectation. This was a score which should occur only one time in 10^{10} (one in ten thousand million) if due to chance. Schmidt attributed their success to some psychic effect.

At the time of my fieldwork in 1972 several positive replications had been informally reported but, for the purposes of this book, the interesting result was the negative replication reported by Beloff and Bate (1971), titled 'An Attempt to Replicate the Schmidt Findings'. The authors introduced their report with a description of Schmidt's results, claiming that:

> ... not only are they of historic importance but constitute one of the most rigorous demonstrations of an ESP effect in the whole parapsychological literature. (p 22)

Referring to the challenge set out by Hansel they wrote:

> On Hansel's own showing, therefore, it seems we must conclude that ESP has now been established and past research has *not* been in vain (p 23, Beloff and Bate's stress).

However, after an elaborate test of 18,650 trials with five subjects and a complex analysis, which searched for any effects outside chance, they were forced to conclude:

> Clearly the attempt to replicate the Schmidt findings has failed.

Far from questioning the existence of the effects demonstrated by Dr Schmidt, however, the authors put the following gloss on their findings:

> We want to make it clear that our own failure to get positive results in no way detracts from Dr Schmidt's success. Why he should succeed and we fail must remain a matter for conjecture. We think it most unlikely that the explanation lies in the difference between our different machines. Nor can it be attributed to the greater length of the Schmidt series since his good subjects show an upward trend almost from the beginning. There remain the imponderables of Dr Schmidt's personal magnetism, his manner, authority etc. that may have inspired his subjects to greater efforts. All we know for the time being is that he found subjects who gave him significant scores, we did not (p 30).

Beloff and Bate, then, hypothesized that the personal qualities of the investigators, or unspecified qualities of the subjects, might be significant variables which accounted for their lack of success. They preferred to count their own experiment as 'incompetent' rather than account Schmidt's work a failure.

The opposite side of the coin was presented in an interview and subsequent correspondence with Professor C.E.M. Hansel. I asked Hansel how he responded to the claims that the Schmidt findings established ESP according to his own criteria. Professor Hansel had reservations regarding the Schmidt apparatus because of its use of integrated circuits which, he remarked, usually need a great deal of 'debugging'. This reservation was backed up by his unwillingness to accept the results of only one experimenter.

> Scientists in general have never trusted investigators. Every result has to be confirmed [correspondence].

> Schmidt should have got other independent investigators to check his subjects to confirm his results. As it is Beloff has repeated his experiment using other subjects and not confirmed the result [correspondence].

> In the case of the Schmidt experiment not all those who have repeated it have got the same result. If half of them get the same result, and half of them don't *one might think that the competence of the investigator was an important variable*. One tends to discover that as soon as one applied rigorous methods in ESP experiments the results disappear. [My stress, my close paraphrase of interview material with Professor C.E.M. Hansel.]

As can be seen, the Schmidt findings presented no special problem for the critic of paranormal phenomena. There were several ways in which they could be explained away. The stressed section in the above paraphrase is particularly interesting. At first glance it would seem that Hansel was making precisely the same point about the competence of the experimenter as was made by Beloff and Bate when they wanted to explain away their negative findings. But what Hansel meant by a competent experimenter is explained in the following extract from correspondence.

> In a particular case, if half the investigators get a result and the other half fail to confirm it, either group could be incompetent or fraudulent. But when the one result [ie, failure] is fully consistent with contemporary scientific ideas and the other radically contradictory to it, the investigators who fail to get a result are more likely to be competent and/or honest.

That is, for Hansel, in this case 'if half the people get the same result' as Schmidt *they* are likely to be incompetent — insufficiently rigorous. For Beloff and Bate (and most other parapsychologists) consistent failure to get results is taken as demonstrating some imponderable personality defect which renders the experimenter incompetent.

Once more one can see the significance of the differing vocabulary and differing conceptual landscape of scientists.[2]

Some propositions confirmed

Propositions Seven and Eight are nicely illustrated by the Backster and Beloff cases. The scientists involved did disagree about which experiments were competently performed; and we can see how the resolution of this debate settled, or would settle, the existence of the phenomena.

Proposition Nine — the coextensiveness of debates over the existence and the nature of phenomena — is also reaffirmed. Backster's unconventional findings and interpretations, like Weber's, had the potential to open out our normal world view. In the main, Backster-type experiments made use of standard techniques. That is, they used control runs, blind judging, automated machinery, statistical analysis and so on. However, when the experimenters came to complain about their critics' deficiencies, they used two types of argument. On the one hand, they pointed to inadequacies of a type which would be understood by, say, any plant physiologist but they also referred to problems which would not be a part of the normal discourse of plant scientists. Comments of this sort will be found among the criticisms of Johnson made in 1972 and in Backster's replies to his AAAS critics. They all refer to features of the *psychic* environment of the experiment.

Some further comments by researchers who claimed positive results make this difference more clear. One of my 1972 respondents, in explaining his success, as compared with the failure of others, to get results agreed that his knowledge of biology was weak but commented:

> I feel, however, that my background in the spiritual world, also my personal experiences with the thing that I'm working with, makes me far more qualified than most people who have Doctors' degrees in biology because I've found these people to be very rigid. I've found them to be very fixed in their area and their training has actually seemed to be more of a restriction in this kind of research than a help.

Marcel Vogel, in discussing the general problems of replicating the experiment, has made the following points in print (Tompkins and Bird, 1974, p 46):

> If they approach the experiment in a mechanistic way, and don't enter into mutual communication with their plants and treat them as friends, they will fail! ... empathy between plant and human is the *key* ... No amount of checking in laboratories is going to prove a thing until the experiments are done by propely trained observers. Spiritual development is indispensable. But this runs counter to the philosophy of many scientists who do not realise that creative experimentation means that the *experimenters must become part of their experiments.*

Believers in psi phenomena are likely to think that this type of factor is the most significant in explaining experimental success or failure. They will use such arguments in trying to determine what are to count as competently performed experiments. Were they to be successful in their arguments, they would have determined certain of the qualities of the phenomenon in question. For example, it would be of such a nature as to be affected by certain subtle psychic influences, and it would not be of such a nature that the mere scientist, trained in a scientific orthodoxy, could hope to discover it.

The division of life, literature and lucre

Separatist tendencies in parapsychology are encouraged by thinking of this sort. Some parapsychologists believe that the existence of psi phenomena has been more than adequately demonstrated in the hundreds of positive experiments that have been reported over the years. They think that it is a waste of time to continue to seek to prove the existence of psi to a sceptical world. They take its existence as proven and try instead to learn more of its characteristics. But in making this decision they have already, as it were, 'constructed' some of its characteristics. For them, not only is psi an everyday phenomenon, but also a frequently reticent phenomenon. Thus they already 'know' something about it. And by observing that, for instance, certain individuals never see it at all, they 'know' more about it still.

Though the waters may have closed over Backster's experiment, there are areas of parapsychology where the conceptual landscape has become very strange through processes of this sort. For example, the failure of experiments has been explained by the retro-active influence of the potential audience of readers of a scientific paper. Though this is an extreme instance, even the routine vocabulary for explaining experimental failure — reasonable though it seems from within the world view of workers in parapsychology — is sufficiently strange to make certain that its users, however well qualified academically, are likely to be shunned by respectable science.

In another paper (Collins and Pinch, 1981), the research trajectories of two physicists who embarked upon parapsychological research and encountered these difficulties has been discussed. One returned to the fold of physics, while the other became more parapsychologist than physicist. These were the only choices available to them because the technical arguments belonging to the two communities had become so different. The divide between these two ways of thinking and the potential impact upon ordinary science, were the technical arguments of parapsychology to be incorporated into normal science, are now very great (Collins and Pinch, 1982).

These tendencies to develop heterodox interpretations of data and of experimental discrepancy are similar to those discussed in the previous chapter. In the case of gravity waves, however, radical interpretations were rapidly suppressed or repressed. In parapsychology, they survive because there is a self contained and separate institutional structure for the heterodox views. Within these institutions what counts as heterodoxy elsewhere is normal. It could be said that what the (or some) parapsychologists have done is to develop their own forms of life. These overlap minimally with ordinary science.

The institutional counterparts of the cognitive divide are the division of literature — different journals are published — and what one might call 'the division of lucre' — parapsychologists rarely compete directly with orthodox science for funding. The careers of people and ideas in ordinary science have their counterparts within the self-contained world of parapsychology. To a Martian, the world of parapsychology would look like a miniature version of its respectable brother. But parapscyhology will never be thought of as proper science on Earth unless it comes to share the institutions and the cognitive life of science.

This conclusion can be expressed as a further proposition:

> *Proposition Ten:* In the long term, phenomena with radical properties can exist only within forms of life and sets of institutions which overlap minimally with science as a whole. Otherwise, either the phenomena, or science, must change.

Using a surrogate phenomenon to break the experimenters' regress in parapsychology

At the end of the last chapter I examined the way that experimenters tried to use calibration as a way of breaking the circle of the experimenters' regress for gravity wave experiments. There is an interesting parallel in parapsychology. To describe this incident I will introduce a third episode of paranormal research, the experiments of Professor John Taylor on the so-called 'Geller-Effect'.

Uri Geller, it will be recalled, claimed to be able to bend metal objects, such as spoons and cutlery, by stroking them gently without exerting manual pressure. Taylor, a theoretical physicist at King's College, London, was one of a number of scientists who experimented on Geller and his emulators — mostly young children — in the mid-1970s (Collins and Pinch, 1981, 1982).

At first Taylor upheld the paranormal claims; he published a popular book (1975) describing various phenomena of this sort. However, after some years of experimentation, he changed his mind and in two jointly-authored papers published in *Nature* in 1978 and

1979 he rejected the possibility of most paranormal phenomena. Subsequently he published another book (1980) containing a longer explanation for his change of mind.

In a much earlier book Taylor (1971) had written that the acceptance of paranormal phenomena *implied* the acceptance of a 'fifth force' in nature: that is, effectively, some 'black box' force that would cause events that could not be explained within the four-force model of conventional physics. He wrote:

> These various paranormal experiences may then be taken as evidence of a new field of force, generated by this new faculty, of a different nature from the four basic ones of the physical world: electromagnetism, gravity, nuclear, and that causing radioactivity (p 221).

Citing a paranormal result, he noted specifically, '. . . *it indicates that this fifth force is certainly not electromagnetic.*'

Thus, before he had much experience of investigating the paranormal himself, Taylor was citing the characteristics of the phenomenon in order to rule out the possibility of an electromagnetic explanation of the paranormal. However, by 1975 he was claiming that electromagnetic radiation was the only known force that could explain paranormal phenomena (Taylor, 1975).

By 1979 he claimed that 'on theoretical grounds the only scientifically feasible explanation would be electromagnetism' (Taylor and Balanovski, 1979, p 631). Given this deduction, a surrogate phenomenon is immediately available for experiments. Instead of looking directly for elusive paranormal phenomena Taylor need only search for electromagnetism. Because electromagnetic forces are well understood, the experimenters' regress does not apply to them in the same way as it applies to the paranormal; so long as the results uphold the orthodox view, no one will question the competence of the experimental procedure. As a matter of fact, Taylor's results were negative, and no one did question the experimental outcome. Thus in 1978 Taylor wrote:

> The quest for EM (electromagnetic) concomitants of ESP is based on our deduction that it is the only known force that could conceivably be involved. . . . In looking for EM signals emitted by people during alleged ESP events we are therefore testing the reality of the corresponding phenomena. There might be no paranormal phenomena at all, so that a search for abnormal EM effects would automatically fail. If we assume that the present evidence for ESP is not firm either way, then its EM characteristics are highly relevant to resolving that problem. If no EM signal were found, this would question the reality of the phenomena whereas suitably strong EM signals would support the claim that ESP effects were occurring. This can be quantified if the sensitivity of subjects to incoming EM radiation is ascertained; for example, sensitivity at least down to the levels of any paranormal emission from others would be

necessary for telepathy to occur. We find no abnormal EM signals during the occurrence of supposed ESP phenomena (Balanovski and Taylor, 1978, p 64, see also Taylor and Balanovski, 1979).

The rest of the papers contain details of quantitative calculations and experimental results at various frequencies of EM radiation that back up this claim.

Taylor's argumentative strategy effectively breaks the circle at the cost of limiting the interpretability of new phenomena. If Taylor's position were to be accepted then henceforward the only forces associated with paranormal phenomena would be forces already well understood. Thus, paranormal forces must either be normal or not exist at all! In this way, Taylor's investigations of the paranormal became part of everyday physics.

Unsurprisingly, the parapsychological community were not impressed by this argument. It is interesting, however, that *Nature* published two papers on the subject without so much as an editorial murmur, whereas anything it has published recently on the paranormal has been surrounded by disclaimers and caveats. Once more, it seems clear that the conceptual universes of these two communities are very different. What seems like a rather silly way of proceeding in one is treated as good sense in the other. Although Taylor's move appears to break the experimenters' regress, it does so only for those who accept the electromagnetic surrogate — a reinterpretation of paranormal forces which can appeal only to hard-line sceptics. Just as in the case of gravitational radiation, only those who are ready to accept a conservative interpretation of the phenomena in question can make use of the surrogate.

Chapters Four and Five show why and how the test of replication fails to work efficiently in disputed areas — the only areas where replication is ever used as a test. They show why and how the murine algorithm, expressed by Popper in Chapter Two, gives rise to sluggish results in the terrestrial computer. The experimenters' regress prevents scientists from agreeing about what counts as a replication. The process of replication described in these chapters contrasts with the TEA-laser work described in Chapter Three; there was never any doubt that the laser could be replicated and never any doubt when it had been replicated. The fact remains that our experience of nearly all natural phenomena is like the experience of laser building; we know that the familiar objects of science are replicable. We have seen some examples of the way that replicability is established, such as Q's confrontational tactics and the uses of surrogate phenomena. We have seen crude conspiracies and refined theorizing used to try to bring a 'closure' to the potentially endless negotiations over whether one phenomenon or another is replicable. It now remains to locate these closure mechanisms in a more general framework and see why some are more likely to succeed than others.

Notes

1. Parapsychology is a marginal field as a whole, yet even within the ranks of parapsychologists Backster was widely distrusted. *The Journal of Parapsychology* (founded by the late J. B. Rhine) had a reputation for refusing to publish articles that claim negative results in parapsychological experiments, yet it published an account of the negative replication of the Backster experiment by Johnson (1972). What is more, Backster was ridiculed by Rhine in an editorial in the journal (Rhine, 1971). Perhaps it is that parapsychologists worry so much about the legitimacy of their discipline at the best of times (Collins and Pinch, 1979) that they found the idea of paranormal sensibilities in the vegetable kingdom too much to countenance.

What is more, Backster was not even a university scientist. He was an enthusiastic layman with an expertise in the field of lie detectors. One might say that Backster was a marginal man's marginal man.

2. Many philosophers of science seem to have missed the way that philosophically generated categories of action change their application in different conceptual landscapes. For example, Popper declares certain hypotheses 'unscientific' because their proponents continually defend them against apparent falsification. This is taken to render them 'unfalsifiable'. Popper's demarcation criterion rests on a failure to see that what may appear *ad hoc* from within one viewpoint will appear to be a brilliant demonstration of experimental expertise (even progressive, Lakatos, 1970) from within another. Suppose Joe Weber had established the existence of gravitational radiation; his defences of his experiments would not now be seen as perversely *ad hoc*, but rather as heroic, after the fashion of a Galileo.

Parapsychologists have the odds somewhat more heavily stacked against them from the start, yet nevertheless a little imagination should make it clear that, in a world where paranormal phenomena were accepted, similar defensive manoeuvres as were made by Backster and Beloff would not have the appearance of desperate ploys (as some readers may be inclined to think of them) but show a wise and proper grasp of the experimental technicalities.

Chapter Six

The Scientist in the Network:
A Sociological Resolution of the Problem
of Inductive Inference

In Chapter One I argued that joint entrenchment in forms of life is the way that conceptual order is maintained. I then examined a number of attempts to establish a conceptual change within science (in Chapters Four and Five). From the scientist's point of view, the establishment of a conceptual change amounts to the widespread acceptance that the corresponding empirical results are replicable. Thus the idea and the practice of replication have been examined. Ten Propositions about experiment have been established along the way. To repeat:

Proposition One: Transfer of skill-like knowledge is capricious.

Proposition Two: Skill-like knowledge travels best (or only) through accomplished practitioners.

Proposition Three: Experimental ability has the character of a skill that can be acquired and developed with practice. Like a skill it cannot be fully explicated, or absolutely established.

Proposition Four: Experimental ability is invisible in its passage and in those who possess it.

Proposition Five: Proper working of the apparatus, parts of the apparatus *and the experimenter* are defined by their ability to take part in producing the proper experimental outcome. Other indicators cannot be found.

Proposition Six: Scientists and others tend to believe in the responsiveness of nature to manipulations directed by sets of algorithm-like instructions. This gives the impression that carrying out experiments is, literally, a formality. This belief, though it may occasionally be suspended at times of difficulty, re-crystallizes catastrophically upon the successful completion of an experiment.

Proposition Seven: When the normal criterion — successful outcome — is not available, scientists disagree about which experiments are competently done.

Proposition Eight: Where there is disagreement about what counts as a competently performed experiment, the ensuing debate is coextensive with the debate about what the proper outcome of the experiment is. The closure of debate about the meaning of competence is the 'discovery' or 'non-discovery' of a new phenomenon.

Proposition Nine: Decisions about the existence of phenomena are coextensive with the 'discovery' of their properties.

Proposition Ten: In the long term, phenomena with radical properties can exist only within forms of life and sets of institutions which overlap minimally with science as a whole. Otherwise, either the phenomena, or science, must change.

Propositions One to Five and Seven are at the root of the experimenters' regress. This, it will be recalled, arises because the skill-like nature of experimentation means that the competence of experimenters and the integrity of experiments can only be ascertained by examining *results*, but the appropriate results can only be known from competently performed experiments, and so forth. Other ways of testing for the competence and integrity of experiments, such as 'tests of tests', turn out to need 'tests of tests of tests' — and so on. Proposition Six shows one reason why the experimenters' regress is a largely invisible feature of scientific practice except in unusual circumstances. Propositions Eight, Nine and Ten are consequences of the other propositions and the regress.

The experimenters' regress has been shown to lie at the heart of the problem of using experimental replication as a test of replicability; the regress prevents us using experiments alone to establish changes in conceptual order. Nevertheless, it has been argued, replicability is a perfectly appropriate criterion for distinguishing the true from the false; replicability is the scientifically institutionalized equivalent of the stability of perception which is, in turn, coextensive with the existence of the corresponding concepts. However, if the replicability of something cannot be clearly revealed in experimental tests it is necessary to turn again to the question of how it and the existence of the corresponding phenomena, are established in practice.

(To put this in terms of the Empirical Programme of Relativism [EPOR]: I have shown that scientists can argue interminably over the meaning and significance of their data and that experiments cannot provide an answer [stage one of EPOR]; I have looked at some of the ways in which scientists bring such arguments to a close in practice [stage two of EPOR]; I now want to look at these 'closures' in the context of the wider network of science and of society [stage three of EPOR]. To do this we need to lift our gaze, from time to time, from the confines of the laboratory.)

The scientist in the network
The first chapter of this book was devoted to 'prizing open common sense reality'. This sort of metaphor — prizing open common sense — has also been used in an entirely different context within the book. In describing Joseph Weber's work I suggested that it was like a volcano which had thrust up through the waters of the everyday practice of physics. That I have found myself using a similar metaphor in two different contexts is more than coincidental. All disturbances of common sense have a similar look about them. They are all attempts at radical cultural innovation.

To understand the context and career of such eruptions we need to understand how a single set of scientific activities relates to scientific practice as a whole. For this we need another metaphor, one also introduced briefly in Chapter One. This is the network — a kind of spider's web of concepts. It is what Barnes (1983a) has referred to as a 'Hesse-net', since the philosopher Mary Hesse (1974) has been its most recent advocate.

A Hesse-net shows how our concepts are joined together in a network of generalizations. Going back to emeralds (Chapter One) we noted that Goodman's explanation of why emeralds were green rather than 'grue' turned on the entrenchment of 'green' in the English language. This, I argued, was not an entirely satisfactory explanation since there are other equally well entrenched colour terms that might be used to describe emeralds. The stability of the greeness of emeralds, I suggested, was due not only to the entrenchment of green but also to the stability or entrenchment of the concept of emerald and of other green things. The stability of *green* and *green things* was mutually reinforcing. It was a matter of joint or multiple entrenchment. The Hesse-net formalizes and generalizes this point. I will borrow the example of the whale from Barnes to describe the way it works.

The concept of 'fish' is linked to concepts such as 'egg-laying creature' and 'creature that cannot breath air' by generalizations. These have the form, 'fish live in water', 'fish lay eggs', 'fish cannot breath air' and so forth. Likewise, the concept 'animal' is linked to other concepts by generalizations such as 'animals live on land', 'animals breath air', 'animals give birth to live young', 'creatures that give birth to live young are animals' and so forth. In the Hesse-net, probabilities are attached to the generalizations. These probabilities express the degree of certainty we have about the way each concept is linked to others; that is, they express the certainty we have about the truth of the inductive generalizations that they embody. If we are really certain that fish live in water then a high probability will be attached to the link between 'fish' and 'creatures that live in water' in the conceptual network. A lower probability might be attached to the link which relates fish to egg-laying creatures. The point is that the whole network is mutually supporting since everything is linked to everything else. But, by virtue of the way that everything is connected, a change in one link might reverberate through the whole of the network.

Barnes imagines a culture in which the concept of fish and the concept of animal are embedded in the net in nearly the same way as they are embedded in our culture. Then he asks what happens when such a culture *first* encounters a whale.

If the whale is included under the concept of fish then this matches such generalizations as fish live in water and fish have fins. But it causes trouble for generalizations such as fish lay eggs and fish cannot breath air. On the other hand, if the whale is included under the concept of animal it matches generalizations about animals bearing live young and breathing air but doesn't fit generalizations such as animals live on land and animals have no fins. Now, there is no absolute sense in which a whale belongs to any category. It is humans who put things in categories. The choice of category in this case is likely to depend upon whether the humans in question would rather cause trouble in the 'fishy' parts of the network or the animal parts. Or, it might be that major difficulties will be avoided by making the whale into a kind of 'anomaly' — a new kind of creature that simply does not fit anywhere for the time being. The latter choice sends no major reverberations through the network.

The network model has a lot to commend it. It shows that all concepts are potentially revisable. It shows that revisions tend to have ramifications elsewhere but they do not have to cause a lot of disruption — the amount they do cause is, to some extent, under control. Finally, though all concepts are potentially revisable, they are not easily revisable all at once — a feature that Hesse stresses.

The point at which I must part company with Hesse's version of the model is the assignment of probabilities to the generalization links in the net. As explained in note 16 of Chapter One, these probabilities do not have useful or recognizable counterparts in practice. Probabilities are altogether too formal a notion to capture the way in which concepts are linked. The network metaphor is exactly right but the links must be something else. *The links are the rules embodied and institutionalized in forms of life.*[1]

The philosopher Peter Winch puts the matter this way:

> A man's social relations with his fellows are permeated with his ideas about reality. Indeed, 'permeated' is hardly a strong enough word: social relations are expressions of ideas about reality. (Winch, 1958, p 23)

We must add that the converse is equally true, that ideas are expressions of social relations. And just as social relations can be described in terms of social networks their cognitive counterparts can be described in terms of the Hesse net. The Hesse net and the network of interactions in society are but two sides of the same coin. To understand each, one must understand both. The form and stability of the Hesse net cannot be understood in terms of interrelated probabilities attached to its links, because the probabilities, and the perceived relationships between them, are themselves expressions of the stability of human relationships. Emerald, then, is linked to green, not by a probability, but by the fact that calling it green is our way of 'going on'. Grass is linked to green in the same way. A proportion of our talk and other actions is linked to these ways of going on in a network of social practices. Sets of practices overlap and it is the overlapping that sends reverberations through the whole system. For example, consider the way that the system would reverberate if certain glittering, clear, carboniferous, ultra-hard stones became part of the concept of the far more common emerald. Emeralds would no longer be green, they would be green or clear. But far more than this would happen. Jewellers lives would change dramatically as the bottom fell out of the diamond market; many small investors would lose a large proportion of their wealth; hearts would be broken; girls would lose their best friends; the title of one of Ian Fleming's books would lose its point; cutting glass would become a much more chancy business; and 'The Emerald Isle' would lose half of its appeal to tourists. These would be the results of a change in the order of concepts. It is better for nearly everyone — except those owning an emerald tiara — that diamonds do not become confused with the type of impure aluminium oxide known as emerald.

Strategy, ambition and the presentation of data

In the emerald example, reverberations spread widely through the network of social relations even though they arose from a relatively 'small' conceptual change.[2] In most of the examples discussed in this book so far, the reverberations did not spread noticeably beyond the institutions of the scientific profession; there are, of course, other scientific changes that have had much more significant social consequences. Inevitably, the extent to which a change, scientific or otherwise, is likely to reverberate through the system as a whole affects the ease with which that change can be brought about. If such a change affects others in ways that they do not like, they are likely to try to resist it.[3] If more are affected, then more resistance can be expected.

Small localized changes are thus usually easier to bring about than large reverberating changes. Bourdieu (1975) suggested that scientists are aware of the possibilities open to them and plan their careers with this in mind. He talked of scientists making decisions between high- and low-risk strategies. They can choose to work in well- understood areas and build a solid career based on gradual incremental advances to knowledge or they can choose to try to make revolutionary contributions which are more likely to be resisted and to lead nowhere. However, if the revolutionary bids are successful, they lead to great rewards. (The rewards are symbolic, not material, and may even be posthumous.) The rewards are great because of the major reverberations and major changes they generate throughout the system. In bringing about major changes in our concepts and ways of doing things the scientist achieves something recognizably important. This something may bring about yet further changes in the wider society.[4]

Bourdieu discussed the differences in the promise of the different fields of science that a novitiate might choose to enter. However, different things can be made out of the same field, indeed, different things can be made out of the *same data*. This is the choice that has faced some of the scientists discussed in this book. We may think of them as spiders sitting in a web of concepts. Their choice turns on how much attention they try to attract to themselves. Given a piece of unwelcome or unexpected data, they may sit quietly and digest it, or ignore it, or they may shake the web until others notice what they have done — and, perhaps, its implied threat. Swallowing quietly will give them a little nourishment but shaking the web may assure them of a glittering future at the risk of disturbing others or creating enemies.

Creating contradiction

This choice can be looked at in a number of ways. An interesting version is to be found in an unpublished paper by Travis called 'Creating Contradiction'. I provide here a simplified, schematic account of a much more complex passage of scientific history.

Travis looked at the controversy over chemical 'memory transfer' (see his 1981). In one series of experiments rats were trained to avoid the dark side of a two-branched alley by an appropriate regimen of rewards and shocks. The results suggested that when the brains of rats which had acquired dark-avoiding tendencies were ground up and injected into untrained rats the latter could be made to develop dark avoiding tendencies more rapidly. It seemed that there was some chemical in the trained rats' brains which 'corresponded' to dark avoidance and which could be transferred by injection into the naive rats; it carried the behavioural tendency with it. This chemical was labelled 'Scotophobin', referring to fear of the dark.

Scotophobin, however, can be thought of in at least two different ways. Scotophobin's effects can be explained if it is thought of as a chemical which brings about a general behavioural disposition. This might work because it makes the creatures fearful of the unknown — as many disorientating drugs do — or it might be a chemical which enhances general learning ability and which builds up in the brain of the trained rats as a result of exposure to so much stimulation. On the other hand, Scotophobin can be thought of as a chemical which carries a kind of memory within it, a memory which specifically views the dark as an unpleasant place to be.

The two different interpretations of Scotophobin have radically different implications. The dispositional drug is just one more discovery among the range of behaviour-modifying chemicals already known, but the 'memory molecule' holds the promise that we might eventually be able to pop down to the pharmacist for a pill containing Greek, or the complete works of Shakespeare. The former interpretation is relatively acceptable within biochemical knowledge and practice, whereas the latter is much more difficult to swallow!

The results of memory transfer experiments became the subject of contention; some groups were able to reproduce them and others were not. Travis says that in these circumstances the proponents of the Scotophobin effect adopted an interpretation with less radical implications in the hope of soothing their critics and making their findings more readily acceptable to the scientific community. But their critics supported the extreme radical interpretation of Scotophobin, arguing that the experiments apparently demonstrated the existence of memory molecules, with all their ludicrous implications. The idea was also said to contradict the 'central dogma of molecular biology'. Thus the critics used this interpretation as a *reductio ad absurdum* of the whole passage of experimentation.

We can easily imagine circumstances in which critics and proponents would have taken precisely the opposite stance. Had proponents wanted to gain some notice and financial support for their endeavours, they might have tried to draw attention to Scotophobin's revolutionary importance by *stressing* that it contradicted dogma and that it promised a complete change in our ideas about learning. Critics might then have argued that Scotophobin was of no fundamental importance since it was only one new behaviour-modifying drug among many. Indeed, Travis reports that such alternative stances were adopted during phases of the long running controversy.

Either way, given the same data, it is possible to rattle the web of concepts more or less, by choosing to create contradiction with what went before, or by presenting the findings as a part of a continuing tradition respectively. It is not quite a matter of individual choice, for others, too, have a say in how findings are to be interpreted.

Prior agreements

Pickering has looked at a number of case studies in which debates were settled by physicists opting to cause minimal disturbance in the network; they preferred to sustain the maximum number of prior alliances. For example, he examined the debate over the claimed discovery of the 'magnetic monopole' (Pickering, 1981).

All known magnets have two inseparable poles. If you cut an ordinary magnet in half each separate half will still have two poles, a north and a south. However, the idea of a monopole particle is not unknown in theoretical physics. In 1975 a group of physicists claimed to have discovered the elusive particle by noting its tracks among other 'cosmic ray' traces in a detector carried aloft by balloon. Others failed to confirm their results and arguments about the competence of different observations ensued. Eventually, as with high fluxes of gravitational radiation, the monopole claim lost credibility and the argument closed. This happened rather more quickly than in the gravitational radiation case.

Pickering concentrates on the way that the closure of the debate was brought about. He argues that all parties proceeded so as to maintain 'prior agreements regarding routine experimental practice and ... theoretical conceptions of the natural world' (p 83). In short, this group of experimenters, even though they found themselves in the position of having to explain odd results, rapidly agreed to maintain the *status quo*. They did not try to press their claims by questioning the wider culture of physics in which routine experimental practice and theoretical conceptions of the natural world are embedded. For example, they did not try to claim that some unknown force prevented others discovering what they had seen or that the theory of cosmic ray detection devices was defective in some way. On the contrary, their acceptance of prior agreements established 'links with theoretical astrophysics and other experimenters' (p 87). The original discoverers preferred to give up their claims and agree that they were mistaken rather than give up these prior agreements. The resolution can be nicely expressed by combining two of Pickering's phrases (from pages 87 and 89): 'The simplicity of the monopole affair derived from the decision of the participants to conduct the debate within a set of socially acceptable conceptualizations of the natural world which were essentially static.'[5]

Externality

Another case has been analysed in a way that makes an important contribution to the picture being developed here. Pinch looked at experiments to detect 'solar neutrinos', which are particles emitted from the centre of the Sun. A proportion of them arrive at the surface of the Earth about eight minutes after they are emitted; if their number could be measured it would yield important clues about the way the Sun works. However, they are very difficult to detect.

To detect solar neutrinos a huge tank of perchlorethylene (dry cleaning fluid) was buried at the bottom of a deep gold mine. Some of the neutrinos passing through the tank ought to strike the chlorine atoms in the fluid and turn them into atoms of radioactive argon. Again, the number of argon atoms is very small and the quantity of argon gas produced at the end of a week or so of exposure is barely measurable. The atoms of argon are detectable only because they are radioactive. Their radioactivity can be measured as they are flushed from the tank with a stream of ordinary argon gas. The radioactivity is detected by delicate electronic devices and the results are recorded on a chart recorder. Just as with gravity waves and plant perception, it is a wiggly line that initially comprises the data.[6]

Pinch (1985) points out that to reach a figure for solar neutrinos emitted by the Sun a number of inferences are needed. However the experiment can be reported in ways that need more or less inferential steps. For example, the experimenters could report that they had done the experiment and seen only a wiggly line on a chart recorder. Or they could say that they had 'seen' such-and-such a number of radioactive argon atoms in their tank in a given time. Or they could say that they had 'seen' such-and-such a number of chlorine atoms converted into argon. In these three cases the number of inferential steps starts at a minimum and increases. Thus to report the existence of a wiggly line is to say nothing about what caused it; to report the existence of a number of radioactive argon atoms is to infer that the radioactive argon was properly flushed, that the radioactivity was properly measured and that the signals were properly amplified or otherwise processed so that they could be recorded; to report the conversion of chlorine atoms implies all this plus the truth of theories about how chlorine can be turned into argon. Of course, to report that a number of *neutrinos* were detected involves still more inference, and to report a '*solar* neutrino flux' requires additional inferences about the nature of the layers of the Sun between the core and the surface and the nature of the space between the Earth and the Sun.

The point is this: there is little future in the scientist's reporting that $100,000 has been spent on sinking an instrumented tank of perchlorethylene in a gold mine to produce a wiggly line. No one is interested in a wiggly line. But, by virtue of its complete vacuousness, the claim that a wiggly line has been seen is unlikely to be contested. It will change no one's life; it will alter no networks of relationships.

Of course, the scientist who has done the experiment is going to report something more. The actual report will be chosen from the range of possibilities that contain more inferential steps; to use Pinch's term, these are more 'external'. The more the externality — the more inferential steps are subsumed — the more interesting the reports become because they touch more on the concerns of others.

Thus the report concerning radioactive argon will be of interest to argon chemists and to those concerned with the design of the counters and amplifiers; the report concerning the conversion of chlorine will, in addition, be of concern to nuclear physicists. The report concerning solar neutrinos will be of concern to all these, and to neutrino physicists, and to solar physicists, and to those interested in the outer layers of the Sun and to experts on the space between the Earth and the Sun.

Still wider groups can be brought in if inferences are made from the 'measured' flux of neutrinos. For example, if the result is interpreted as revealing something about the state of the thermonuclear processes at the heart of the Sun, and this is taken to have implications for historical changes in its energy output, then there are further implications for theories of terrestrial geology and evolution. (As a matter of fact, the results have been interpreted along these lines.)

The scientists, then, are faced with a choice (albeit, a highly constrained choice); at what level of inference, or externality, do they report their results? The more inferences they make the more interesting the results are to a wider and wider audience — the more they rattle the spider's web of concepts, as it were. But, if the results are not likely to preserve everyone's 'socially acceptable conceptualizations of the natural world', then the more inferences they make, the more bits of taken-for-granted reality they are threatening, and the more trouble they are going to cause.

Though the choice of making more inferential steps does not in itself 'create contradiction', it certainly does draw attention to any contradictions that can be found in the data. It would be a choice opposite to that made by many of the scientists that Pickering studied. For instance, the monopole scientists chose to break the chain of inference which went from their data to the existence of the monopole by admitting a 'mistake'.

Gravitational radiation and parapsychology revisited

This analysis applies equally to some passages of the gravitational radiation controversy. For a time those who were debating Weber's claims were prepared to grant that he was seeing more than some statistical artifact; they were prepared to grant that he was seeing *coincidences* between his detectors. These, they felt, must be due to something other than gravity waves since no-one but he could see them. This move, in shortening the inferential chain, removed cosmologists and relativity theorists from the argument as it removed the great 'cosmic energy drain' implied by Weber's claims. Under these circumstances such groups would no longer find their taken for granted worlds disturbed; they would not feel any need to resist the existence of 'mere coincidences'; they would not see Weber as creating contradictions. The cost, of course, is that the coincidences become somewhat less interesting. Correlated television signals or some other mundane phenomenon are not likely to change scientists' lives very much nor win any prizes.

In this example, like the previous ones, the extent to which a claim is hard to make acceptable increases as it threatens to prize open the conceptual world of the scientists whose 'homes' lie in remote parts of the conceptual web. The analysis is continuous with the very extreme cases reported in Chapter Five. Parapsychology threatens too much to too many to be easily acceptable. That is why its more uncompromising proponents are forced to live in a world of their own (see Proposition Ten). Their web of concepts, *and the coextensive social network*, has fewer connections with the main network of science than do most scientific fields.

Thus the same data can be made to reverberate more or less and it can also be made to reverberate in different directions: it can be associated with different vectors, as it were. To extend the spider's web analogy, it is as though different spokes of the web can be rattled so as to send the reverberations toward different parts of the social and conceptual universe.

Professor John Taylor's ultimate rejection of a fifth force to explain paranormal phenomena and his tight embrace of the electromagnetic hypothesis could well be described, in Pickering's words as preserving the maximum number of 'prior agreements'; but these were agreements with physicists. He chose to sacrifice all his agreements with parapsychologists! What is more, if he had found the enormous fluxes of electromagnetic radiation that would be required to explain 'paranormal spoon bending', then he would effectively have shifted the locus of the problem away from physics toward biology. That part of the web of concepts to do with physics would have remained quiet, whereas biologists would have found themselves having to explain how human beings can generate fluxes of electromagnetic radiation sufficient to soften metal. There would have been few agreements preserved with biologists; biologists were not, perhaps, so important a 'reference group' for Taylor.

Likewise, Weber's reluctant acceptance of electrostatic calibration shifted the locus of debate from the physics of the interaction between his antennae and gravity waves toward the arena of cosmology. Henceforward his results had to be explained by odd pulse shapes implying 'pathological' cosmic scenarios, which met with resistance from cosmologists. But remember that back in 1972 there was talk of a 'fifth force', or even psychic forces, as explanations of the growing discrepancies in antenna performance. If these explanations had been pressed then there would have been resistance throughout nearly the whole of the web, balanced only by new alliances forged with the almost disconnected parapsychologists. Weber might have found himself cut off (see also Collins and Pinch, 1981). The possibilities open to Weber — and most of them were considered at one time or another — represented almost the whole range from psychic radicalism to timid 'unexplained coincidences'; and all with the same 'data'. To think of these moves merely as *ad hoc* attempts to rescue a hypothesis is to miss the dimension concerned with the different cognitive communities to which different interpretations appeal or pose threats.

The quality of these local communities within the web was well expressed in a talk given by theoretical physicist Kip Thorne at the 1978 meeting of the American Association for the Advancement of Science in Washington D.C. Thorne reviewed progress in the gravity wave affair. He constructed a graph showing expected frequency of occurrence of gravitational wave events against their magnitude. The less energetic events might be expected to appear more often than larger events. Less energetic events originate in more distant cosmic catastrophes and there is simply more universe as one looks further. However, only the most energetic events could be detected with Weber-type antennae. Thorne made clear the assumptions upon which his calculations rested. He said:

> ...I have ... [shown] ... the maximum possible strength of sources that you might have on certain *cherished beliefs* that theorists hold dear to themselves about conservation of mass, about non-beaming of gravitational waves — which might be a false cherished belief — and about the strength of gravitational wave emission from individual sources. [my stress]

Then in response to a question from the audience he explained himself this way:

> ...I and several associates at Cal Tech have made a list of things that we hold very dear. We can actually exhibit, then, scenarios that would lead to bursts [of gravity waves] that were that strong, with that kind of frequency [so that they could be reconciled with Weber's claims]. But they are scenarios that an astrophysicist would say are really pretty wild. Nevertheless they don't violate any of our cherished beliefs. Now your cherished beliefs might be different to mine, so you might have different curves there.

Some astrophysicists, of course, cherish deeply the absolute impossibility of Thorne's 'wild scenarios'.

Yet another set of evidence shows that physicists' cherished beliefs are different to those of parapsychologists. During my fieldwork on parapsychology certain respondents volunteered comments on Weber's experiment without prompting; these were to the effect that they thought that psychokinesis represented a parsimonious explanation of the difference between Weber's results and those of his critics. In later interviews I deliberately asked eleven parapsychologists to comment on the Weber affair. Eight of the eleven thought that Weber's results were a prime candidate for a psychokinetic explanation. In other words they thought that Weber's results could be accounted for by his intense *desire* to find the waves; this desire had influenced his immensely delicate apparatus.

One respondent remarked:

> Weber did the most sophisticated PK [psychokinesis] experiment ever run ... The signals are real ... and are just about the right level for a PK experiment with a good PK subject ...

Thus cherished beliefs are different in different parts of science. The preservation of cherished beliefs or prior alliances cannot in itself determine the outcome of a debate even when it is the prevailing motive. The question of *whose beliefs* are to be preserved still remains. The perceived importance of the different communities in the web to members of other communities explains a lot about the way that debates in science proceed. In Pickering's study of the charm/colour debate (Note 5), an alliance with a group of mathematicians seems to have been influential. In Pinch's (1977) study of Von Neumann's proof, it is the perceived status of mathematics that seems to explain what happened in the most parsimonious fashion. On the other hand, alliances with parapsychologists seem to be given up with great readiness. (There are some exceptions, see Collins and Pinch, 1981.) Only when debate is kept to a low level, or a low degree of 'externality', is it likely that a relatively unified community can be addressed.

Science and the more distant
regions of the network
The examples of scientific controversy examined in detail in this book did not have reverberations that spread noticeably beyond the institutions of professional science. There have been no modern studies of contemporary pure science which show the ramifications of influence outside the scientific community.[7] A number of case studies of 'the influence of social interests' on older passages of science have been completed. The extended study by Mackenzie (1981) on the development of the correlation coefficient within statistics is a good example. Another interesting study is Shapin's (1979) examination of disputes over phrenology in Edinburgh in the early decades of the nineteenth century. Shapin traced the interests of certain classes in supporting phrenology. If innate character could be read from the bumps of the head, not only would this put intellectual 'outsiders' in a position to make judgments on human nature which were previously thought to be the preserve of professionals, it also suggested the development of a new social order — one that would accord better with physiognomical

indicators of ability. A political programme therefore went along with the endorsement of phrenology.

The study shows in detail the way that competing political preferences led, through different perceptions of the shape of the outside of the head, to different perceptions of the cavities within the bones of the skull — the phrenologists required that the outside of the skull should be parallel to the inside so that the shape of the brain was reproduced — to different perceptions of the shape of the surface of the brain right through to different perceptions of the structure of brain material itself — the phrenologists saw the surface linked by fibres to the spinal cord, whereas anti-phrenologists saw brain substance as more homogeneous so that separate organs corresponding to aspects of character could not be distinguished. All these differences can be seen in the drawings of the brain reproduced in the texts of the competing authors. Shapin's study links the most intimate details of brain anatomy to distant parts of the network concerned with class and status in Edinburgh. There is no reason in principle why modern scientific controversies should not reverberate to distant parts of the social network in the same way. Such links and influences are not at all inconsistent with what has been argued in the rest of the book. Such a model links social structure to the laboratory bench.

The core set: social contingency with methodological propriety

Scientists know most about those parts of the conceptual web that make up their own discipline. To some extent their views are formed by their background and their aims will be mediated by their picture of the corresponding area of the web. The picture is first developed during scientists' training and continues to develop as a result of their relations with colleagues and through their continued work. Allies and critics in a controversy prefer to preserve different sets of alliances in the wider network depending on their own backgrounds and training. The arguments and attitudes of different allies and critics will be affected by their perceptions of their place in the web and their ambitions and strategies. The set of allies and enemies in the core of a controversy are not necessarily bound to each other by social ties or membership of common institutions. Some members of this set may be intent on destroying an interpretation of the universe upon which others have staked their careers, their academic credibility and perhaps their whole social identity. If these enemies interact, it is likely to be only in the context of the particular passing debate. This set of persons does not necessarily act like a 'group'. They are bound only by their close, if differing, interests in the controversy's outcome. I refer to such a set of allies and enemies as a 'core set'.[8] The controversies described in Chapters Four and Five were acted out by core sets.

Core sets certify new knowledge. From the outside they appear to be simply the 'group of scientists' who are investigating a potentially novel feature of the universe. But there is a theoretical tension in the idea of a core set that needs to be properly understood. In the previous chapters we have seen that the activities of core set members do not accord with the conventional image of 'scientific' investigation. The knowledge which emerges from a core set is the outcome of an argument that may have taken many forms not normally viewed as belonging to science. All these 'negotiating tactics', I have suggested, are attempts to break the experimenters' regress. Some 'non-scientific' tactics *must* be employed because the resources of experiment alone are insufficient. In the absence of an algorithmic recipe for proper replication of an experiment, these tactics are ways of trying to establish what is to count as 'going on in the same way' in the future.

Nevertheless, the outcome of these negotiations, that is, certified knowledge, is in every way 'proper scientific knowledge'. It is replicable knowledge. Once the controversy is concluded, this knowledge is seen to have been generated by a procedure which embodies all the methodological proprieties of science. To look for something better than this is to try to grasp a shadow. Scientists do not act dishonourably when they engage in the debates typical of core sets; there is nothing else for them to do if a debate is ever to be settled and if new knowledge is ever to emerge from the dispute. There is no realm of ideal scientific behaviour. Such a realm — the canonical model of science — exists only in our imaginations.

Core sets funnel all of their competing scientists' ambitions and favoured alliances and produce scientifically certified knowledge at the end. These competing ambitions and alliances represent the influence or 'feedback' from the rest of the web of concepts and therefore the rest of our social institutions. Different scientists in the core set will be differentially aware and differentially interested in the troubles and pressures in remote areas that would be caused by one outcome rather than another. Their arguments and strategies will be formed accordingly.

The ramifications, as we have seen, are not necessarily limited to the boundaries of the scientific community. The links to industrial and political interests are sometimes subtle (as for the cases discussed in this book) and sometimes, as in the case of, say, the debate over the development of new nuclear power stations, they are obvious. The core set 'launders' all these 'non-scientific' influences and 'non-scientific' debating tactics. It renders them invisible because, when the debate is over, all that is left is the conclusion that one result was replicable and one was not; one set of experiments was competently done by one set of experts while the other — which produced the non-replicable results — was not. The core set 'funnels in' social interests, turns them into 'non-scientific' negotiating tactics and uses them to manufacture certified knowledge. If one looks very closely, one can see how the outcome of core set debates is affected by these 'socially contingent' factors, but one can also see how the output is nevertheless what will henceforward be proper knowledge. *The core set gives methodological propriety to social contingency.*

The private nature of core sets

In general, core sets are private, and that makes it hard for most of us to understand what science is. Only the very few scientists who have been involved in a major dispute have substantial experience of core sets; even for these scientists the experience is likely to be a fleeting one. Doubtless all scientists know that there are difficulties in experimental work, but the significance of this is seldom understood. We saw a paradigm case of this sort of experience in Chapter Three. Bob Harrison began to be puzzled about the proper way to proceed in a number of areas related to the TEA-laser, but as soon as he made it work he crystallized this experience into a series of failures on *his* part rather than drawing a more general conclusion. This point was summarized in Proposition Six.

As for the rest of us, our experience of core sets is nil. Hardly anyone has extended experience of what it is like to produce new scientific knowledge out of a controversial area. It is interesting that we feel, nevertheless, that we have a fairly good grasp of what the scientific method comprises. We learn this from doing some science at school, from watching demonstrations on the television and from reading about science in books and newspapers. Even most philosophers of science work with some version of the 'canonical model'. All these sources stress the infallibility of experimentally generated knowledge. And yet, if we consider which aspects of the world we are most certain about, it turns out that we have no direct experimental experience of them at all. Consider, for example, how certain we are about the truth of relativity, yet few readers of this book will ever have designed or conducted an experiment concerned with Einstein's theory. The point applies just as much to scientists as to the lay reader. The irony is that knowledge at a distance feels more certain than knowledge that has just been generated. *The degree of certainty which is ascribed to knowledge increases catastrophically as it crosses the core set boundary in both space and time.* Even for core set members it is very hard to recapture the uncertainty of the time of creation once the debate is closed and the correct way of going on has been crystallized into the new scientific institutions. As for those who have no first hand experience of core set work, it is almost impossible to know what the creation was like: to know what it is like to put the ship into the bottle. This point can be summarized as another proposition:

> *Proposition Eleven:* 'Distance lends enchantment': the more distant in social space or time is the locus of creation of knowledge the more certain it is.[9]

For scientific culture, the mediating role of the core set, its laundering of 'illegitimate social interest', and its transubstantiation of social contingency into methodological propriety, along with its privacy, explain the paradox of reification.

Order and changing order: the sociological
resolution of the problem of induction

Throughout these chapters I have talked about the same thing using a variety of vocabularies. I have talked about the unspecifiability of rules in open systems, the foundations of taken-for-granted reality and forms of life, the problem of developing artificial intelligence, the impenetrability of tacit knowledge, the opaque 'murine rules' for replicating the 'same' experiment and the experimenters' regress. I have tried to show how each of these are but different facets or consequences of the same underlying problem, the problem of inductive inference.

I have looked at various attempts to provide algorithmic procedures for solving all of these versions of the problem. I started with attempts to explain greenness and the proper continuation of '2,4,6,8'; I went on to look at some attempts to provide an analytic theory of replication and attempts to provide criteria of demarcation at the various levels of the imaginary murine sorting procedure; I noted scientists' attempts to break the experimenters' regress by reference to any number of factors ranging from theoretical presuppositions — 'cherished beliefs' in Kip Thorne's terms — to procedures such as the use of surrogate phenomena and calibration. All these attempts were bound to fail if they were intended to provide sets of formal rules which transcend the conventions of the society in which they are embedded. As we have argued, rules are only rules by virtue of social conventions: they *are* social conventions. When new rules are being formed, or when old rules are being applied in new ways, it is precisely new social conventions that are being established. How does this process present itself to the individual?

Early on in a controversy the members of the core set, and any others who have an opinion, think of the experiments in a dichotomous manner. They think of experiments which give support to their views as competently performed, and vice versa. These points of view can be summarized in a table:

Scientists' Views Regarding the Competence of Experimenters and the Integrity of Their Experiments

		Scientist believes in phenomenon under investigation	
		Yes	No
Experiment finds results	Yes	1. Competent	2. Not competent
consonant with phenomenon	No	3. Not competent	4. Competent

In 1972, Joseph Weber would have put his own experiment in box 1 and those of his critics in box 3. His critics initially saw the same experiments as belonging in boxes 2 and 4 respectively. In 1972 Cleve Backster and his critics would have classed their experiments in the same way.

Beloff and Bate (Chapter Five) are interesting because they did not trust their own experimental abilities when their work produced negative findings, so they put their own experiment into box 3. Hansel, on the other hand, put Beloff and Bate's experiment into box 4 since he thought their failure to find the phenomenon of psychokinesis showed how well they did the work!

A little later, but while core set members are still disagreeing with one another about whether some result has been properly replicated, things change somewhat. Thus by 1975, in the case of gravity waves, critics described the majority of negative experiments as inadequate, though all reserved box 4 for at least one or more negative experiment. This finding is repeated in the Backster study; in that case, by 1975, the Johnson work had been dropped from the competent experiment categories by critics as well as believers. Every box still had occupants but some of the negative experiments had dropped right off the diagram by this slightly later stage.

It seems that while there are only a few negative experimental results available, those predisposed to disbelieve in the existence of a phenomenon will treat all these as competent for they *need* experimental legitimation for their negative views. All such experiments go into box 4 under these circumstances. Later, when there are many experimental 'refutations' to choose from, critics can afford to abandon some of them. They can even demonstrate their fair-mindedness by pointing out the loopholes! But this will not empty box 4. Likewise, when there are few positive experiments available they all go into box 1, but if there were more, some of these would be cast overboard too.

At a still later stage the crystallization of the core set's work amounts to the collapse of one side or the other of the table. If the work is vindicated then boxes 2 and 4 disappear from viable discourse. If it is agreed that the phenomenon does not exist then it is boxes 1 and 3 that go. There is no longer a form of life that will support the institutions pertaining to the absent boxes. Only isolated individuals can now order their concepts and language around high fluxes of gravity waves or the secret life of plants, and a private language is no language at all.

This is where Bob Harrison and the TEA-laser fit in. Harrison worked from the beginning in a 'crystallized' or thoroughly institutionalized area. Clearly he classed his own work in box 3 when the laser did not lase and box 1 when it did. Boxes 2 and 4 were simply not available to him or anyone else. By the time Harrison was building TEA-lasers, no sideways re-categorizations were possible. The days of the possibility of sideways transfers of TEA-laser experiments finished in about 1969. After that, all competent experiments were in box 1 — they produced the phenomenon — and any experiment that did not produce the phenomenon was in box 3. The institutionalization of a new piece of science is the closing off of one side of the table and the shutting down of the possibility of sideways re-categorizations. It also amounts to the collapse of the sorting problem discussed in Chapter Two. While there would be sorting problems for the imaginary mice in the cases of gravity waves and parapsychology, there would be none in the TEA-laser case.[10] This was because the answer was known at the outset. Quite simply, the sole defining criterion for a successful experiment was that it produced the expected result. All other activities were rejected instantly and without further consideration. Thus similarity and difference were easily recognised. The embedding culture had institutions that corresponded to the notions that a transversely excited tube of gas at atmospheric pressure was only to be counted as a copy of a TEA-laser if it could lase. If the device lased then it must have passed through every sorting stage. If it did not lase, then it must certainly have fallen at one or more of the hurdles.

That is *the sociological resolution of the problem of inductive inference*. We perceive regularity and order because any perception of irregularity in an institutionalized rule is translated by ourselves and others as a fault in the perceiver or in some other part of the train of perception. Thus we arrange all our perceptions into boxes in just the same way as Bob Harrison arranged his perception of his own experiment. If we do not care to do this, then others will not be able to communicate with us; they will treat us as though we spoke a private language. Our regular judgments of similarity and difference are made with no more difficulty than attends the categorization of experimental activities once the proper outcome has been defined. The open system quality of these judgments disappears when the appropriate outcome is known. After this there is only one proper continuation to the series. It is not the regularity of the world that imposes itself on our senses but the regularity of our institutionalized beliefs that imposes itself on the world. We adjust our minds until we perceive no fault in normality. This is the meaning for the individual of the joint entrenchment of a concept. It is why our perceptual ships stay in their bottles.

The process is robust because of the way single concepts are linked to others through the overlapping of forms of life. Others will correct or ignore us if we make mistakes. The lasting quality of order arises from the resistance of the network — the massed spiders in the conceptual web. The locus of order is society.

This is a picture of social and conceptual order, but if there is to be substantial change then new ways of proceeding must be invented and sustained. But in earlier sections of this chapter we have seen how easily contradictions can be initially created. A potential scientific revolution can be read into any trivial mistake. Thus the origin of creativity in itself is not an interesting problem. The interesting thing is the origin of *successful* creativity and the conditions for its success. What is involved in making major changes in the conceptual web?

The individual and society

Individuals should be thought of as the sum of the forms of life in which they play a role. For most purposes an individual's thoughts *qua individual* are of no interest. The most useful way of thinking about the goals of members of the core set is by thinking of those members as 'delegates' from the disciplines or other social and cognitive institutions which form their background. When a sociologist interviews a scientist he or she is really talking to a set of forms of life. As it happens, the only way to find out about forms of life is to talk to the individuals who share them; it would be easier were it otherwise.

Nevertheless, it is only individuals who can provide the material for conceptual change. It requires that someone be prepared to risk a new way of 'going on'. It is individuals' abilities to choose to go on in the 'wrong way' that makes them creative. It is their ability to do the equivalent of playing Awkward Student (playing 'Awkward Scientist') that provides the means for change.

But, just as with Awkward Student, a new continuation cannot be just any continuation. It must fit with some part of the network. It must be more than the equivalent of a rude noise or complete nonsense if it is ever to be more than a private language.[11] This provides constraints upon the possibilities of creative success. An individual act of creation is worth nothing unless it can become institutionalized. Again, the wider network and the wider society provide the conditions for the success of some new institutions but not others. 'Man proposes but society disposes', one might say. The problem of the individual and society, or microscopic and macroscopic explanation in the social sciences, is a matter of the detailed interrelationship of proposal and disposal. Thus wholly structural or wholly microscopic explanations nest into one another. Structural explanations are adequate in their own terms but take for granted the steps that link them to individual activity. Microscopic explanations always refer, explicitly or implicitly, to the influence of a cultural/structural context even if they do not include an analysis of it.[12]

Awkward Scientists, if their efforts are to be met with success, need allies within and outside the core set. Allies outside the core set can offer material help such as financial and professional support, publishing outlets, publicity and so forth. This is the stuff of the 'politics of science'; since it is not the subject of this book I touch on it only in the postscript. Allies within the core set can offer two sorts of help. They can help scientists with mental work and they can help them 'legitimate' the heterodox new ways of going on.

The day to day maintenance of a position through argument is hard work. Creating new arguments which defend a novel stance and which are not entirely unreasonable is hard work. One can get the feel of this by practising Awkward Student. The arguments generated must be 'plausible'. That is to say it must be shown that they do not entail the wholesale otherthrow of enormous sections of the network. Much of what went before must look the same if the new ideas are to have any chance of success. This is true even of scientific revolutions. Most of the rest of our social institutions remain intact even during a 'revolution'; at least they must appear to remain intact at the time. Distant retrospective analyses may discern more profound changes throughout conceptual life, but at the time some sense of continuity is vital. Wholesale rejection of science merely invites total exclusion from scientific discourse in the way that a 'silly answer' invites exclusion from the game of Awkward Student. The difficulty in producing the arguments only arises because most social institutions have to seem to be preserved.[13]

As we saw with Awkward Student, allies can help in the construction of arguments which maintain an awkward position but which still preserve enough of existing social institutions so as not to seem totally silly. Allies can convey ideas privately so that the Awkward Scientist can use them. But there is a much more useful thing that allies can do. They can *act* — and this includes acts of speech — as though the ideas are reasonable. Thus can they 'create plausibility' for a new idea.

In Chapter Four it was argued that the cosmic scenarios proposed by Weber to explain why only his detector could see high fluxes of gravity waves were 'implausible'. But we also saw that the plausibility of arguments that turned on the unsuitability of electrostatic forces as a calibrating surrogate could have been increased if a better 'facsimile' of gravity waves, such as the rotating bar, had been available at the time. When the rotating bar calibrator was *suggested* the difference between electrostatic forces and gravity waves became a little more apparent. 'Mere talk' of alternate forms of calibration made Weber's argument a little less outrageous.

Harvey (1981) saw similar things happening in experiments to do with quantum theory; a scientist's mere willingness to test a bizarre hypothesis rendered it far more plausible.[14]

In both the calibration of gravity waves and the quantum theory tests, simply inventing an experimental design made certain hypotheses into something that needed to be eliminated experimentally rather than by default. Conceptual possibility in these cases depended not on theory but, in a very palpable sense, on activity — or at least proposed activity. Thus the realm of ideas is not circumscribed by the limits of human thought but by the limits of what people *do and say* in society. The actions described maintained an opening for a more widespread change in physics' form of life than the first, and at that point natural, interpretations of the experiments would have allowed.[15]

Thus, while inventing and whispering new arguments represents valuable help from an ally, if the ally *uses* the idea it is much better.[16] The use might comprise practical application but even speaking the arguments out loud rather than whispering them has a substantial effect. In saying new things the scientist is taking part in a new institution. There is not a lot of difference between saying and doing for both are public ways of demonstrating that there is a new set of rules that can be followed — a new way of going on in the same way.

Correspondingly, the first ploy (and an extremely effective one) of a sensible and determined scientific critic, is to ignore the contentious claim. Even to criticize an idea in a devastating way is to start to bring about its institutionalization. Only if the proponent scientist is him or herself a very effective publicist and advocate is there nothing to be lost by an early frontal attack.

Changing order
The recipe for changing order starts with an individual who is

prepared to put forward an intepretation of data which has the potential to create some contradictions and reverberate through the social and conceptual web. This interpretation, however, must be such as to seem to preserve most existing institutions; the new account must not appear to be completely unreasonable. In this, the Awkward Scientist can use the help of colleagues. As individuals, scientists see a battle between competing accounts of the value of experiments with competing outcomes (as represented in the table displayed earlier in this chapter). The scientist who wants to press an awkward claim is trying to maintain the legitimacy of the accounts that allow the corresponding experiments to be seen as competently performed. Remember that the experimenters' regress prevents an 'objective' solution so the matter is in flux until the new convention becomes crystallized. The proponents of the radical, or awkward, interpretation can keep the possibility alive by acts of speech or other actions so as to make them seem plausible. Correspondingly, critics' best early strategy is silence. Any other tactic or exercise of power is usable and used within the core set in order to try to bring about a favourable 'closure' of the debate. These are the individual's contributions to changing and maintaining order.

Critics' and allies' positions are formed in a large part by their backgrounds and alliances within the social and corresponding conceptual web. This is one way in which the larger society feeds back upon the actions and choices of the scientists at the centre of a dispute. Through the same mechanism, interests and influences also bear upon the continuing course of the argument. The differential success of the competing positions within the core set is partly explained by the way that members are tied into the web. Material influence as well as matters of plausibility are significant. Whether the web was shaken more or less hard, or in one direction rather than another, makes permanent change more or less easy to bring about since it attracts the attention of different groups. In this way, the 'same data' may be effected more or less, and by more or less powerful groups within the scientific and the wider community. These are the contributions of the wider society to stability and change.

The mediating institution — if such a loose and transient set of persons should be called an 'institution' — is the core set. In the core set the actions of individuals and the influences of the wider web are melded together. The ferment is apparent when a core set is the object of close scrutiny, but the core set's more typical privacy is what allows this ferment to give rise to expert knowledge. As far as the general and scientific publics are concerned, the crystallization of the work of a core set — which coincides with its disappearance — is the end of a passage of creative work by the appropriately qualified experts; they demonstrate the replicability of one or other set of claims. If there are still dissenting voices then they are seen to belong in a quite palpable way to scientific non-persons; these are persons who, because of some individual pathology, are unable to accept the 'truth'. The ferment of creation, which was never public in the first place, is then virtually irrecoverable. As for the individual scientist who has experience of controversy, for all but a few (who either keep quiet or are expelled from regular scientific society) the experience is one of temporary aberrations in the norms of scientific deportment. Once the scientific truth is known it is forgotten that non-experimental and 'non-scientific' negotiating tactics were necessary if closure was to be attained. The magic of the core set lies in the way it uses anything to make a scientific fact yet also renders all the ingredients invisible to all but the very determined investigator.

The aim of this work has been to cast light on questions of the grounds of knowledge and on broader questions of social and conceptual order. Through the examination of science I have tried to show how individuals create the potential for change, how other individuals can help or hinder this process, how these efforts are embedded in the wider society, how the wider society is the locus of conceptual order and how facts are made fact-like in spite of their human creators. Finally, and this is a corollary of the last point, I have tried to explain why the version of events described here is so little known. If we always had the image of the core set before our eyes then there would be no facts. The privacy of creation is what maintains its sanctity and its power.

Notes

1. It is not clear that Barnes would not want to say something similar about the links in the network. Bloor's (1983) exposition of Wittgenstein accords closely with what is being argued here.

2. But it is not an entirely fanciful example. Consider what turns on gradations of colour in South Africa. The distinctions between white and coloured in South Africa rest on concepts that are not available to the rest of the world. Consider again the sub-division between Jew and Aryan created in Nazi Germany. The deaths of millions were brought about in part because a conceptual distinction was established which enabled a sub-set of a nation's population to be distinguished from their fellows. When Denmark was occupied by the Nazis the local inhabitants (led by their king) resisted the application of the concept 'Jew' in a very direct way. When the Jews were ordered to sew yellow stars to their clothes as a distinguishing mark, the rest of the population did likewise. Thus the yellow star could not become a perceptual counterpart of the concept of Jew.

3. The idea of resistance to scientific discovery is not new. See Barber (1961) for an excellent early discussion.

4. Myers has recently completed two studies which illustrate some of the points in the discussion which follows. In the first study biologists submitted papers for publication. They were forced to claim less and less originality for their ideas as they

responded to the comments of critical referees. To get the articles published, the biologists had to make them read as much more continuous with biological tradition and much less original than they first thought them to be. In a second, equally fascinating, study Myers shows how similar pressures force biologists to adjust and re-adjust grant applications in response to external comments. Grant applications which look too original will not get funded (see Myers, 1985a and b).

For an interesting discussion of science as a marketing exercise see Peter and Olson (1983).

For a fascinating account of Pasteur's struggles to translate the concepts of the farm and the laboratory into one another as he developed his vaccine against anthrax, see Latour (1983).

5. This sort of explanation is used by Pickering in two other case studies from the same area (1980, 1981b, 1984). These are the debate over the proper theoretical analysis of the nature of the basic building blocks of matter — charm versus colour — and the debate over the existence of another elusive particle — the free quark. In both cases he again explains the closure of debate by the formation of alliances with established groups in the scientific community; in both cases the 'logic' of the situation would have allowed radical arguments to have kept the matter open for longer.

Pickering stresses, quite rightly, that the contexts within which these debates take place are subject to change and that, therefore, preference for context stability cannot be the complete explanation of the construction of knowledge. He suggests that the appropriate place to look for an understanding of the context is within the 'theoretical conceptions' of a discipline. There is certainly some truth in this if by theoretical conceptions he means concepts in the sense that this term has been used throughout this book — the organisation of perception, rooted in social institutions, whether the things that are perceived are quarks or emeralds. If he refers specifically to the theories of physics then his idea is too limiting. The concepts that make up the scientific world are formed as much, or more, by manipulating it experimentally as by theorizing about it.

6. The story of the difficulties and disputes over the detection of solar neutrinos is itself a fascinating one (Pinch, 1981 and forthcoming) but here I will look only at the treatment to be found in Pinch (1985).

7. Indeed, one of the latest trends in empirical studies of contemporary science is a focus on more and more detailed and microscopic aspects of life within the laboratory (eg, see Knorr-Cetina, 1981, 1983; Lynch et al, 1983). Interesting though these studies are, their narrow field location makes it difficult for them to take account of the wide social base of legitimate knowledge (Collins, 1983c, but see Knorr-Cetina and Cicourel, 1981). Interestingly, one of the pioneers of laboratory studies Latour, (eg, see Latour and Woolgar, 1979) has worked hard to demonstrate the links between the laboratory and the wider society but he has done this for a historical episode in the work of Pasteur (Latour, 1983).

It is interesting to speculate on the lack of similar studies of contemporary pure science. It may be that scientific institutions have become more autonomous so that the social network between science and the wider society is now sparse. I think it far more likely that it is a matter of not being able to 'see the wood for the trees' in very recent scientific history.

There are, of course, many interesting studies of the relationship between modern *applied* science and social interests: for example, see Studer and Chubin, 1980; Robbins and Johnston, 1976; Gillespie, Eva and Johnston, 1979; Nowotny, 1977; Mazur, 1981; Nelkin, 1975, 1978, 1979; Petersen and Markle 1979; Markle and Petersen, 1980. However, it is difficult to draw from these the very general point that I want to make in this book, since studies of science with obvious political connotations can always be treated as 'special cases' of distortion (Chubin, 1982; Collins, 1982a). To some extent, of course, the same applies to historical studies.

8. The core set is likely to be small in size and range from perhaps a couple of scientists to fifty. For further discussion of this concept, see Collins (1979).

The interactions of the members of an 'ideal' core set might be represented on a diagram in the following way:

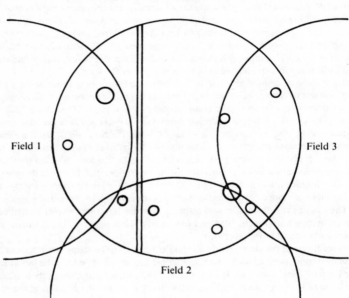

The Way Core Set Members Might Split
Among Different Fields of a Science

An 'Idealised' Core Set Within the Network of Science

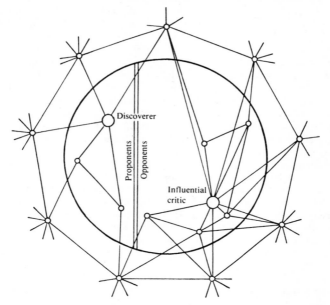

Note: the lack of direct relations between proponents and critics.

9. A similar point was made by Ludwick Fleck (1979, p 113).

Latour and Woolgar (1979) talk about the way that an idea becomes real as it moves from the seat of its creation. They call this 'splitting'. When the new phenomenon — in their study a new drug — splits from the set of activities and writings that initially comprise it, its reified form gives unity to what went before. All that work, it turns out, was directed toward making the new drug. Before the drug existed, or if the drug were never to become a thing in itself, there would be nothing that linked all those laboratory activities as part of the 'same' piece of science.

The sudden and catastrophic crystallization of certainty — even for the participants themselves, is what leaves one very distrustful of retrospective accounts of the closure of controversies such as are cited by some historians. Some of these give the impression that the facts 'spoke for themselves' on many occasions.

10. The greater the uncertainty — in other words, the more widespread the potential threat to the network — the more sorting levels are intractable to us humans. Thus in the parapsychology case study there were potential disagreements all the way up. As a matter of fact, it was the problems at the very top level which provided the greatest difficulties of all in the Backster case and the most ingenious criticisms of his work. (The top level has to do with identifying activities that are anything to do with the phenomenon in question.) As the debate went on, it became clear that, since we are continually surrounded by instances of death, it is impossible to know when the plants would be suffering the emotional stimulation that Backster needed to isolate.

In Backster's argument with Gasteiger about control drops, it emerged that tap water was disqualified as a control drop substance because it contained micro-organisms.

These were killed by dropping them into hot water just as the lives of the brine shrimps were extinguished. However, once that had been noticed, other objections became reasonable. For example, every time the urinal in the laboratory block was automatically flushed with disinfected water the plants might well feel sympathy for the germs destroyed! Then consider the experimenter's footsteps across the laboratory floor. Were ants being crushed? Then consider the massive battles for survival being fought out by the insect and microbe populations of the floors, walls and ceilings and on the person of the experimenter himself. And why restrict consideration to the laboratory? What about remote locations? Is there an inverse square law for death? Backster could, of course, have invoked some kind of 'ad hoc' limiting hypotheses such as an inverse square law or a lower limit on the size of an organism for which a plant could feel, but these began to look implausible after the tap water argument had been deployed.

Turning to gravitational radiation, the sorting difficulties did not extend so far. Thus there were arguments about how to assess the aggregate sum of the results (Level Seven); there were arguments about whether one result or another was positive or negative (Level Six); and there were arguments about whether some experiment was competently performed (Level Five). However, there were no arguments at Level Three about whether the identity of the experimenter was appropriate (except that Joe Weber's self replications were not taken to account for a great deal) and there were certainly no arguments about whether what was going on fell within the purview of science (Level Two). Finally there was never any question concerning mundane activities that might really be manifestations of gravitational radiation (Level One). (All these levels would open up were the psychokinetic interpretation of Weber's work to be pressed.)

The most interesting case is the replication of TEA-lasers. Here there was no difficulty in sorting at any stage! Here the components of the mouse-machine worked in perfect crystalline harmony.

11. Gooding (forthcoming) has looked at the way individual scientists create 'construals' of new phenomena so as to make it possible for themselves and others to grasp the ideas. The construals can be based on homely constructions of everyday objects. (See Nickles, 1980 and Brannigan, 1981 for other interesting discussions of discovery.)

12. The Empirical Programme of Relativism is microscopic at stage one, but refers explicitly to structure at stage three. This chapter shows how stage one links to stage three through the mediation of the core set at stage two.

13. This is not an argument for social and scientific conservatism; revolutions involve radical reinterpretation, not replacement, of what we know.

14. Harvey examined the experiments performed in the 1970s to test the possibility of 'hidden variables'. These would allow that a causal order underlies the random world of the quantum event. Quantum theory states that the behaviour of the tiniest constituents of matter can only be described probabilistically. That is, it insists that we can never have an explanation of why, say, a radioactive particle splits at one moment rather than another. We can only say that in a certain period of time there is a certain likelihood that it will split, or that given a large collection of such particles it is almost a certainty that a proportion will split during the course of a given interval. This is an extremely successful theory. It is the orthodox view in physics, but a residual dissatisfaction remains, encapsulated in Einstein's famous aphorism, 'God does not play dice.' Thus, over the years attempts have been made to develop so-called 'hidden variable' theories that would ground the apparently irreducibly random events in a more deterministic foundation. After many years in which it was (incorrectly) argued from first principles that such a thing was impossible (see Pinch, 1977) an experimental test was discovered for at least one class of hidden variable theories.

After the first series of 'non-locality' experiments had been completed it was conceded that they did not support the hidden variable interpretation. Still, a bizarre loophole called the 'timing hypothesis' would allow the hidden variable interpretation to be maintained.

This hypothesis amounted to the idea that separated measuring devices were communicating with each other faster than the speed of light. At first it was not taken to be a serious possibility, but nevertheless after the completion of the first phase of experiments a French scientist named Alain Aspect thought of a way of testing the timing hypothesis and set about building the apparatus to do it. Aspect wanted to close the last loophole in the argument. However, just as in the cases discussed earlier, Aspect's mere *willingness* to test the possibility increased its plausibility; effectively he opened the loophole, at least for a time. This happened even before he had done the experiment.

15. For a very interesting discussion of the way concepts of this sort develop see Barnes (1983b).

16. 'Ask not for the meaning but for the use.' This Wittgensteinian sentiment underlies the theory. Pickering's work can be seen in this way too (1980, 1984). He relates the demise of colour theory in particle physics to the ready use of techniques pertaining to its competitor 'charm' by a group of mathematicians. Pickering talks of the 'dynamics of practice'. Hacking (1983) suggests that it is the *use* of new ideas and findings in distantly related fields and experiments that establishes them.

Postscript: Science as Expertise

Two models of science and their implications

If privacy is the condition of the creation and stability of facts, why invade it? A chink of light is necessary only because, in its attempt to combat less enlightened forces, science has taken, and been granted, the mantle of infallibility. This is dangerous, not only for other institutions, but for science itself. Thus, the purpose of this book is not to reveal 'flaws' in scientific knowledge but the 'artisanship' of its construction. The object of this postscript is to indicate the wider significance of such a perspective.

New and timorous sciences, such as the whole range of social sciences, have tried to develop by apeing what they believe to be the method of the natural sciences — particularly physics. A false picture of scientific method also informs thought and practice wherever science and its findings are administered, discussed, used, or cited in argument outside the research front of science itself. The false picture of scientific method arises from what I call the 'algorithmical model' of learning, teaching, communication and practice.

The algorithmical model encourages the view that formal communication can carry a complete recipe for experiment with all that follows. It encourages the view that the formalized accounts of scientific work found in the journals are complete accounts. The reader, the model suggests, has been a 'virtual witness' of scientists' activities and can see the validity of the procedures and findings.[1] 'Scientific' has become a synonym for 'certain'; scientists' views are *authoritative*. This leads, in the lay mind, to the application of 'flip-flop logic' where science is concerned. Each authoritative pronouncement of science, it is believed, is either wholly true or the outcome of incompetence, distortion or fraud such as to make it wholly false. To question the results of a passage of scientific work amounts to an accusation. There is no middle way.

By contrast the starting point for the model of science developed here — the 'enculturational model' — is the acquisition of skill as opposed to formal instruction. The locus of knowledge is not the written word or symbol but the community of expert practitioners (this includes communities of theorists). Individuals' knowledge must be acquired by contact with the relevant community rather than by transferring programmes of instruction. Scientists are to be seen as expert consultants rather than infallible authorities. Varying, even contradictory, expert opinions are to be expected occasionally under these circumstances; the cause of variation is not necessarily incompetence, bias or fraud. The flip-flop logic is no longer persuasive. Variation of expert opinion is to be treated as natural and ordinary rather than as an area of eradicable weakness or bias. So far as absolute certainty is concerned, it is an impression given by social and temporal distance from the seat of creation of knowledge.[2]

The enculturational model's first implication for the 'methods' of science is that the 'soft' sciences must not take as their favoured model canonical but false versions of the hard sciences. To make unfavourable comparisons between natural and other sciences in terms of their 'scientific' appurtenances is to misunderstand the nature of the enterprise. However politically expedient in the short term, to act 'scientifically' for its own sake — meaning to act out the imaginary canonical/algorithmical model of science — is misguided.[3]

Rather, the notion of a community of skills provides new sets of questions for science policy. For example, the skilful communities which support research of every kind need study. These communities comprise the unglamorous, unexciting, infrastructure of science. The danger is that in times of financial stringency, money-saving policies might reduce the extent of the prosaic infrastructure so as to press forward the more glamorous, if 'safe', aspects of scientific life. This is not a viable policy for the long term. The other end of the scientific spectrum needs protection too. The ratio of orderliness to innovation — of the normal phase and the extraordinary phase — needs to be kept in balance. In times of financial stringency the risky, extraordinary phase of science is likely to suffer disproportionately as compared with the pressing but predictable parts of the normal phase. Conservative impulses have the channels of influence too heavily loaded in their favour (Collins 1983b). The progress and development of scientific expertise can only continue if the conditions for *changing order* are not entirely destroyed. While no one can say what the correct balance is, we may be sure that a society which provides no possibility for radical conceptual development is one in which short term practical concerns have gained too great an ascendancy.

Science education
The skilful community is also the image of science that needs to inform science teaching. The practical science lesson, however much its underlying ethos is one of free discovery, is and must be 'stage managed'. In an interesting article Atkinson and Delamont (1977) looked at the way the 'discoveries' of school pupils are managed so that the 'right' results emerge from a session of discovery work. The descriptions of scientists' work in earlier chapters should make it clear that the 'right' result is the last thing that school children left to themselves will produce. The discovery method must be a sham.

Stage management involves, in a subtle way, shifting the responsibility for lack of success in discovery on to the student. This is how you and I and Bob Harrison begin to develop the propensities to blame ourselves for any perceived recalcitrance of nature. It is necessary for us to blame ourselves if we are to learn appropriate skills and learn to 'see' the world like scientists. An impression of chaos is the alternative; science cannot be learned that way. But the only picture of the physical universe that can emerge from the way science must be taught to potential scientists is one of rigid certainty. It is a picture in which nature has never displayed any inconsistency, only man or his artifacts.

One of the consequences is that a student who does go on to become a research scientist is traumatized by the first experience of real research. Might the PhD drop-out rate be decreased if some expectation of trouble could be built into the curriculum a little earlier on?

For the future *citizen* the model of science and the natural world that is developed through normal scientific teaching is positively dangerous for democracy and for the long-term future of science itself. The model allows the citizen only two responses to science: either awe at science's authority along with a total acceptance of scientists' *ex cathedra* statements, or rejection — the incomprehending anti-science reaction. This is the citizen's interpretation of flip-flop logic. Where scientists' *ex cathedra* statements are found wanting — as they inevitably will be from time to time — then the most likely reaction is disillusion and distrust (Collins and Shapin, 1984).

But in the classroom, where several groups of students are working on similar projects at the same time, we already have a miniature version of a core set. By the end of the lesson all experiments produced by the class must fit the left hand side of the table found in Chapter Six. The reduction of the initial disorder to order — all of the children accepting the teacher's version of the correct outcome and all interpreting their own work in a way which is consistent with the teacher's version — is the resolution of a core set controversy writ small. Might it be possible to take time, every now and again, to examine the very stage-managing process that goes into the successful practical science lesson? The process of resolution is there to be examined and described. A few hours set aside from science training could be used to pull out from the practical science lesson the real method of the construction of order in the natural world.[4] Such a practical *social science* lesson could prepare the future scientist for the world of research and help the future citizen to understand and contribute through the ballot box to expert scientific and technological decision making.

Understanding the deconstruction of facts

As argued in the previous chapter, perception of certainty is a matter of distance from the scene of the crystallization both in time and 'social space'. Certainty increases because the details of the social processes that went into the creation of certainty become invisible. There is a 'catastrophic' change as the boundary of the core set is crossed. Thenceforward certainty is maintained by continued representations of the data in the style of facts. But the recipe for the construction of certainty also provides the recipe for its deconstruction. By focusing attention on, and painstakingly re-describing experiments in all their contingent details, facts can sometimes be deconstructed again. This tactic can be seen at work in debates about parapsychology (Collins and Pinch, 1982). What the scientific paper leaves out is what the critic brings back in. Irrespective of whether the critic describes 'truly disqualifying' acts of clumsiness or incompetence, or irrelevant details, the mere act of describing an experiment as a piece of ordinary life reduces its power to convince.

The core set, it will be remembered, acted as a funnel for social contingencies and interests. It sucked them in and 'purified' them into proper knowledge. Detailed redescription of core set activity reverses the flow through the funnel. Contingencies and interests are separated out once more. Honest redescription of the process of creation, provided it is sufficiently detailed, can start to get the ship back out of the bottle.

Much of what happens where science and technology enter the public arena is best understood in terms of the construction and deconstruction of facts. The scientist-expert wants to engender the maximum facticity; the critic wants to reverse the flow and reveal the 'hidden interests' and contingencies. Perhaps another expert wants to replace one set of facts with another; the first must be deconstructed to make way. The proper way to look at this is as an argument without a definitive solution. Flip-flop logic, though it may remove the anxiety associated with uncertainty, does not capture these shades and nuances of construction and deconstruction.

The tension between the flip-flop model and the construction-deconstruction model is manifested, for example, in the British legal system. A court case is one of the rare occasions on which it is in someone's interests to try to deconstruct some piece of very ordinary scientific work, work that backs up expert forensic evidence, for example. Again, this is a place where the citizen — as defendant, or member of a jury, or judge — is likely to be asked to consider the meaning of expert opinion.

In the British legal system the state, through the medium of the police and Home Office laboratories, handles most of the forensic science. It is the state which prosecutes; thus the defence usually does not have as much access to the scientific evidence as does the prosecution. This would not matter if data were as neutral as the conventional picture of science makes them out to be. If the same data were the inevitable precipitate of the experimental algorithm *whoever* carried it through, then data produced by the state would as likely speak for the defence as for the prosecution. At the very worst, the problem would turn on matters of interpretation, not evidence.

The ironies of this arrangement were nicely illustrated by the head of the Home Office Forensic Science Laboratories, Margaret Pereira, in a TV programme interview.[5] The presenter asked why defence scientists were not invited to participate in the process of forensic investigations.

> Pereira: There have been occasions when defence lawyers have visited forensic science labs to see what is available [pause] but when it comes to advising them how to defend a specific case, I think that that is really asking too much — is it not?
> Presenter: Only if you are on the side of the Crown.
> Pereira: I disagree with that because the point is that these [pause] data are available and what you are asking is for us to tell them how to distort them almost, because that is quite a lot of what is done. The game is different for the prosecution and the defence.

Here the contradiction between the accepted view of scientific data, and Pereira's practical experience of what can be done by the determined critic is nicely brought out. If they are data, they should speak with like voice to any member of the scientific community whether working for the defence or for the prosecution. If they are the crystallized opinions of experts, then those opinions might crystallize differently for defence and prosecution. In that case disagreement over the soundness and quality of data should be *expected*; it should be the norm that the 'same' materials can be turned into different end products. Were defence and opposition scientists given equal access to the materials, then again, we would have a core set. Judges and lawyers, as well as their clients, should understand the nature of the scientific contest (see also Oteri, Weinberg and Pinales, 1973).

Technology and public inquiries

Technology is open to the same analysis as science proper and its influence impinges upon the citizen more frequently. Luckily, precisely because technological decisions are so much more a part of everyone's day-to-day life, the message is easier to grasp. Put simply, the argument is that what turn out to be optimum technological solutions to problems are only obviously the optimum solutions in retrospect. At the time those solutions are being 'crystallized' their socially contingent quality is far more clear; in the case of technology there is a more obvious route between larger social interests and eventual solutions.[6]

The technological equivalent of the criminal court's forensic science is the evidence of experts at public enquiries. A public inquiry is like a trial in that great energy is expended by various parties in trying to reveal the contingent foundation of judgements that others claim to be based on certain technical authority.

A microcosm of the problems faced by inquiries of this sort is in risk assessment. How are the risks — of, say, nuclear power plants — to be assessed? There are a number of ways of providing an apparently authoritative numerical solution to this problem. For example, if the probability of failure of each component of a plant is known from experience of past performance in related environments, the probability of overall failure can be calculated by some sort of aggregation exercise. The aggregation exercise might look at the likelihood of cascades of component failures that would lead to a catastrophe. But, of course, such figures are open to reinterpretation. The 'objective' calculation of the risks in a nuclear plant depend on assumptions about the continued regular behaviour of the plant's components. But each new plant is a new environment and the total environment in which the plant is set is an open system. Thus, not all potential configurations of the plant and potential relationships with its environment can be foreseen. The number that emerges from a risk calculation is the 'crystallization' of an open system argument. It is no surprise that different experts will come up with different numbers. It is no surprise that there are different opinions as to whether a meaningful number can be reached at all.[7]

Thus, different experts will produce different opinions regarding the quantity and justifiability of risks and likewise different experts will produce different opinions about the other more formal technological aspects of the workings of nuclear plants. It is this, according to Wynne (1982), who examined the Windscale Inquiry of 1977, that was incomprehensible to Justice Parker, who presided over it.

Imbued with the conventional empirical model of scientific rationality Parker treated discussion of the interests underlying expert viewpoints, not as revealing the relevant social factors within the scientific debate but as accusations of personal dishonesty. 'I have no doubt as to the integrity of those concerned in all of [the controlling authorities] and I regard the attacks made upon them as without foundation' (Parker quoted in Wynne, p 131). Within the view that facts speak for themselves to unbiased observers, disagreement over the meaning of facts can only be interpreted as accusations of dishonesty. This is the flip-flop model of scientific knowledge.

Such a view may be appropriate *within* science, where it acts as the separate motor for the supporters of each of the competing viewpoints and as a perspective that 'seals' and stabilizes the eventual crystallization of one, but it is inadequate when technical decisions have to be made in the political arena. The flip-flop model raises the level of accusation and counter-accusation, leads inexorably to the discrediting of one set of scientific experts or another, and in the long term will lead to loss of confidence in the whole scientific enterprise. A loss of confidence in the scientific enterprise is a disaster that we cannot afford. For all its fallibility, science is the best institution for generating knowledge about the natural world that we have.

The politics of science; science as politics

The model of the establishment of scientific facts put forward in this book dissolves the boundary between science and technology and the rest of society in two ways. First, it points to the continuity of the networks of social relationships within the scientific professions with networks in society as a whole. Second, it points to the analogy between cultural production in science and all other forms of social and conceptual innovation.

Networks ramify continuously so that reverberations induced within science have their effects outside just as influences from outside the scientific professions feed back into science proper. Science and technology are affected in quite straightforward ways by the political climate. Within scientific institutions the formal instruments of legitimacy such as the journal and professional appointments can be offered or withheld from researchers. Funding can be controlled. Publicity, which can prevent a new idea being stillborn through its critics' silence, and can have more indirect influence on the career of an idea, can also be controlled.

More mundane political forces also play their part in different ways. Some scientific ideas seem more sympathetic to certain political notions than others. For example, Shapin (1979) showed the link between Edinburgh politics and phrenology and MacKenzie (1981) showed how the eugenics movement affected the development of statistics. These influences are readily understood because they make up the everyday analysis of the history of cultural and political change. If scientific change *is* social change, then the same categories of explanation must apply — though this book is not the place to analyse them in detail.[8]

We can understand scientific change by looking at the politics of social change but, if science is a representative example of cultural activity, we ought equally to be able to learn about social change by looking at change in science. I hope other sociologists and political scientists will be able to use this and other modern studies of science to illuminate more general problems. I will end by offering three indications of the small-scale parallels between scientific and political culture.

First, acts of speech and silence have their political counterparts. Radical political movements will try to engender forceful opposition to make themselves visible. If the state attacks a new movement then that movement becomes a movement worthy of attack; its status is defined by the very attack itself. The problem for the state is when to attack. The recent history of neo-fascist movements exemplifies the dilemma. Such movements cannot be entirely ignored, yet would a head-on attack on the embryonic fascist movements in contemporary Britain actually serve to give them legitimacy? This political dilemma is the counterpart of the dilemma in science. It is only worth attacking a 'bad' idea once it has gained the degree of recognition that makes it certain that the attack itself cannot increase it further.

Another parallel is in the maintenance of the locus of authority. The strength of the core set as the mediator of proper knowledge is vividly brought out by Q's need to do an experiment in order to make certain of his membership (discussed in Chapter Four). How is this boundary maintained and how is it broken down? Professional institutions such as the journal and scientific societies help to restrict the actual and potential membership. Specialized language helps to maintain the boundaries around new subjects (Shapin, 1984). The fact that membership demands experimental skill, and the corresponding financial resources, maintains boundaries around all but very new and cheap subjects (see also Gieryn, 1983). The way that authority is maintained in science is a model for the more self-conscious boundary maintenance of all professions.

Finally, there is a political analogue to the processes described under the heading 'Creating Contradiction' in Chapter Four. Terrorist violence, like scientific data, can be interpreted in radically different ways. Those interested in maintaining the *status quo* interpret acts of violence as 'contradicting' civilized behaviour. On the other hand, terrorists try to press an interpretation of regular behaviour as itself violently repressive. For example, in a poverty stricken country with a high infant mortality rate, the deaths of babies may be presented as acts of violence by the ruling powers against the population at large. Thus, terrorists try to legitimize their violence by presenting it as the same as the violence of the state. The state decries violence as a radical, and unacceptable, innovation.

Authority and expertise

Why is it that the view of science as a human product is so difficult to maintain? I have already argued that the privacy of core sets provides the particular answer in the case of science. Scientific training forces the experience of nature's caprices to be interpreted as personal failure thus understressing the human contribution to the achievement of conceptual order. But, we can learn to live with the enculturational model of science.

We understand the fallibility and interests of financial advisors, lawyers, politicians, art and literary critics, doctors, builders, car mechanics, and travel agents without concluding that they are not more expert in their areas than we ourselves. Neither anarchy nor nihilism follows from the recognition of the human basis of expertise; instead comes the recognition that there is no magical escape from the pangs of uncertainty that underlie our decisions.

Professional scientists are the experts to whom we must turn when we want to know about the natural world. Science, however, is not a profession that can take from our shoulders the burden of political, legal, moral and technological decision making. It can only offer the best advice that there is to be had. To ask for more than this is to risk widespread disillusion with science with all its devastating consequences.

Notes

1. Shapin (1984) traces the origin of this idea to Robert Boyle. He suggests that Boyle tried to describe his experiments in such a way as to make the reader feel that he had actually witnessed the experiment itself. He calls this 'virtual witnessing'. But scientists' attempts to repeat experiments by using only written sources, as described in this book, suggest that, if the reader of a modern article felt as though virtual witnessing had taken place, it would be an illusion brought about by the propensity described in Chapter Three as Proposition Six and by Proposition Eleven — distance lends enchantment — since all such attempts ended in failure. (See also Shapin and Schaffer, 1985.)

2. Polanyi was the first to stress the skilful nature of science. Oddly enough he argues from this to the necessary authority of science (eg, 1962). He argues that though knowledge is fallible it is only those who are part of the republic of science who can take meaningful part in its production. Authority must therefore remain in their hands. I argue, from the same starting point, that the community of science cannot always be removed from the arena of politics. These are occasions when expert (as opposed to authoritative) knowledge forms a part of public decision making.

3. See also Overington (1979) and Feyerabend (1975).

4. Robert Millikan rejected some of the initial runs of his famous oil drop experiment because they seemed to reveal fractional charges. He was concerned to produce a good clean result, and preferred to 'doctor' the data slightly in order to produce it. See Holton (1978) for a fascinating discussion.

The matter has become more complicated recently because 'free quarks' would reveal themselves as a fractional charge — one-third or two-thirds of the charge on the electron. William Fairbank and his collaborators claimed to have discovered such charges in the early 1970s (eg, see Pickering, 1981b).

Since the oil drop experiment is often done in schools and universities, I would like to collect sets of physics undergraduates' laboratory notebooks for the few years surrounding the publicization of Fairbank's apparent discovery of the free fractional charge. I would then compare students' reports of their Millikan oil drop experiments before and after this 'discovery'. My hunch is that the students' notebooks would reveal a higher proportion of fractional charges shortly after Fairbank's claim than they did before. This would give us an interesting measure of the degree of self-imposed stage management in the the practical work of physics undergraduates.

5. BBC TV *Panorama*, 'The Whole Truth', 18 April 1983.

6. For two interesting papers on the way that the geological assessment of crude oil reserves serves the interests of different parties, see Bowden (1985) and Dennis (1985). For the relationship between the sociologies of scientific and technological knowledge, see Pinch and Bijker (1984). For other references, see note 7 of Chapter Six.

7. For an interesting discussion of risk assessment, see Critchley (1978). For an interesting account of the way that different views on technological risk are related to different social positions, see Cotgrove (1982).

8. Frankel (1976) has argued that scientific revolutions can be analysed as a species of social revolution. His analysis of the history of optics uses political categories.

9. I owe this section to Graham Cox.

Methodological Appendix

Fieldwork

The main body of material described in this book was gathered on various field trips between the summer of 1971 and March 1979:

1. In the summer of 1971 British physicists building or trying to build TEA-lasers were visited and interviewed.

2. In 1972 British scientists working on gravitational radiation and parapsychology were visited and interviewed.

3. In the autumn of 1972 North American scientists working on TEA-lasers, gravitational radiation and parapsychology were visited and interviewed.

4. In late 1974/early 1975 I was able to work closely with a physicist at Bath University (Bob Harrison) while he built a TEA-laser.

5. In the autumn of 1975 North American scientists working on gravitational radiation and parapsychology were interviewed a second time. British scientists were interviewed a second time around the same date, as were German scientists working on gravity waves.

6. Material which is not central to this book but which is referred to was collected in Britain and the USA in 1976 and 1977 (see Collins and Pinch, 1982). This material concerns the parapsychological area known as 'paranormal metal bending' or 'spoon bending'.

7. In March 1979 I was again able to work closely with Bob Harrison for a few days while he resolved the teething troubles on a second TEA-laser that he had built. This time the work was done at Heriot-Watt University in Edinburgh.

The work referred to under 4, 6 and 7 involved participation in experiments with working scientists. Some other experimental work was done in parapsychology (see Chapter Five) but otherwise field trips comprised interviews with relevant scientists. Such interviews were extended free-ranging discussions centred on technical questioning. In all cases these discussions were tape recorded. Edited transcripts of the tape recordings were prepared later. In all cases I tried to interview all the scientists involved in the scientific work in question. However, logistical considerations kept me to Britain and North America except for the one trip to West Germany referred to under 5.

Representativeness

Are these studies 'representative'? The second two studies — gravitational radiation and parapsychology — are unrepresentative in that being studies of controversial science they stand for a tiny minority of scientific activity as a whole. What makes this an attractive area for examination is that much of what is hidden in ordinary science is brought out in controversy. Scientists, in the main, are not prone to analysing their own procedures. So long as results are steadily produced the results can speak for themselves. Only when trouble of one sort or another arises does reflectiveness about method develop. For example, it is most unlikely that the parapsychologists' discussion of the meaning of replication, presented in Chapter Two, would have been generated by a set of more ordinary scientists. It is only people like parapsychologists (or sociologists!) who are sufficiently tortured about their own procedures to feel the need to analyse them in such detail.

Some critics, in conversation, have argued that the statistical unrepresentativeness of controversy goes along with a more serious problem. They feel that the study of controversy cannot reveal the mechanisms pertaining to the maintenance of consensus in more normal areas, nor the mechanisms of more gentle change in such areas. I have always found this a very odd argument. It seems to me that the resolution of a controversy is precisely the establishment of a new consensus, so that understanding one automatically yields an understanding of the other. No doubt, however, there is a lot to be said for examining smoother waters too. Indeed, once the right questions have been generated, and we have become sensitive to the turmoil beneath the surface through our examination of controversy, then normal science looks the most remarkable phenomenon of all. The TEA-laser study is precisely an examination of such smooth waters.

There may be other respects in which the case studies are untypical. Here we engage with the philosophical matter of the meaning of 'the same'. The meaning of 'representative' is just another version of this, as Goodman (1978, pps 134ff) makes clear. I see nothing that makes these studies untypical at the outset, but the only answer to this sort of criticism is to challenge critics to repeat the work using 'the same' (!) methods. So far everyone who has looked in the same way has come back with a broadly similar story, and the critics have tended to be those who prefer to use methods which rely on accounts generated after disagreements have been settled. An account need be generated only very shortly after a disagreement has been settled for it to have an entirely different quality to one produced while the account-giver is in a state of less than complete certainty. This point — Proposition Six — is discussed further in Chapter Six.

Replication and isomorphism

As mentioned at the end of Chapter One scientists rarely feel sufficiently exercised about knowledge claims to want to turn the theoretical idea of the replicability into a *test* of reality. Thus another unusual feature of the studies reported here is that in every case deliberate attention was given to replication. The motive varied, however. In both the gravity wave and the parapsychology cases replications were carried out as tests of previous observations. In the TEA-laser case the sole aim was to make a working laser as quickly and easily as possible.

For the reasons explained in the section on the analytic theory of replication in Chapter Two, and at the beginning of Chapter Four, scientists building gravity wave detectors did not try to build *isomorphic* copies of Weber's apparatus. (By isomorphic I imply similar in every knowable, and readily achievable, significant respect.) Their aim was to build as good a gravity wave detector as they could. (The consequences of this for the efficacy of their work as a *test* is explained in Chapter Four.) Likewise , in parapsychology, the importance of isomorphism of experimental design is a point of disagreement; isomorphism is not an obvious route to take. These case studies would be 'cleaner' if the aim of the scientists had been to construct isomorphic experiments; my major point, about the variability of scientists' perceptions of what counts as repetition of their work, would then be more clearly made. As it is, a layer of complication overlays the conceptual issue of what is seen to count as 'the same'. (The German gravity wave experiment, and the Beloff and Bate psychokinesis experiment, were, however, both fairly isomorphic in aim.)

For reasons which ought by now to be readily understood, it would be virtually impossible to find a case of replication being used as a test in 'normal science'; and yet the comparative dimension is important. Thus, to find a case of replication in normal science, we are bound to look at a non-test situation such as the TEA-laser. It turns out that the non-test quality is precisely what makes this case interestingly different. It is only because the expected results of the experiment are uncontroversial that it is possible to replicate uncontroversially.

Further, since the TEA-laser scientists expected rewards, not for building a laser, nor for advancing TEA-laser design, but from the work they intended to do with the laser beam once the device was working, they had no interest in building anything other than isomorphic designs. An exact copy, in so far as significant details of design were concerned, was likely to be the most efficient way of realizing their aims. Thus, the laser study represents a 'clean' case of isomorphic replication. (Benjamin Matalon of the Université de Paris VIII has done interesting studies of replication style and frequency.)

Methodological presuppositions
In 1971 at the outset of these studies on lasers the project was not envisaged as a study of replication but of knowledge transfer. The intention was to explore knowledge transfer in a manner informed by ideas drawn from the philosophy of social science and the history of science. The most important idea drawn from the philosophy of social science was that actors are to be understood as acting within a 'form of life' (Winch, 1958; Wittgenstein, 1953). This idea was taken to have its counterpart, in the history of science, in the notion of 'paradigm' (Kuhn, 1962). With these ideas in mind the 'communication network' of TEA-laser builders was explored.

The study of communication networks is well established in the sociology of science and in information science. Many such studies have been carried out, but nearly all treat communication in much the same way. The research techniques they use are suitable for the exploration of the transmission of *information* where information is taken to be organized into discrete visible and articulateable units. Even where information scientists discuss 'informal communication' it is taken that informality is a property of the *medium* not the message; in principle the contents of informal communication could be written down and transferred through, as it were, a 'journal of informal communication' if it were not for questions of logistics, convenience and secrecy.

For communication visualized in this way, questionnaire responses, mutual citations and so forth are reasonable indicators of the transfer of information because the transfer is readily visible to all parties.

Such techniques are not appropriate if communication is viewed differently, however. For example, if it is taken that an important component of scientists' knowledge is 'tacit knowledge' (Polanyi, 1958) then the transfer of that knowledge to the scientist is likely to have been as invisible as the knowledge itself. More generally, if an actor's knowledge comprises his 'form of life', and a scientist's knowledge comprises his paradigm, then the way that they came by that knowledge, or even elements within that knowledge, is unlikely to be properly investigated through means designed to explore information.

The techniques I used to explore the transfer of knowledge in the case of TEA-laser building were adopted to cope with knowledge as a form of life or knowledge as a paradigm that must internalized. Thus, the case study approach seemed appropriate; one could trace, in detail, the transfer of a piece of knowledge through a number of scientists; one could watch this process of transfer in action, rather than rely on scientists' reports which would inevitably refer only to what was visible to them. Also, and this is a crucial point for the whole argument of the book, there was readily available an indicator of the successful transfer of knowledge; when scientist X's laser 'lased', he had the knowledge to build a laser; when it didn't, he hadn't. Thus reliance on scientists' beliefs about their own knowledge — which could only be reliable where that knowledge was articulateable — was avoided.

The methododological preference in this book is for participant observation as a way of developing the native skills that form a base for discussion. I have tried to use extended interviews as a substitute for this where full participation was impossible.

Participant observation goes along with an interpretative approach to sociology (see Collins, 1979, 1983a, 1984a). In terms of the long running argument about the identity of the social and natural sciences (eg, Yearley, 1984) the significance of such a methodological posture has changed. In the 1960s, it would be correct to say that interpretivists believed that sociology should not be treated as a science because, at that time, to be 'scientific' implied some version of positivist/behaviourist methodology in the social sciences. With the growth of the sociology of scientific knowledge, and some recent trends in philosophy of science, notions of scientific methodology have become more pluralistic. A more useful definition of science now turns on the claim of replicability (as explained toward the end of Chapter One). There is not the slightest reason why observations, made from an interpretivist/participant observer viewpoint, should not be replicable by those who are prepared to acquire the relevant native skills. The methodology of this book should therefore be seen as one among the range of methodologies belonging to the sciences.

'Meta-methodological' presuppositions

I differ in approach to some modern writers in the matter of presuppositions. For instance, Mary Hesse and Barry Barnes want to reserve more of their explanations of knowledge for the fundamental 'physics and physiology' of situations than I find necessary. The phrase 'physics and physiology' is Hesse's. She argues, contra Popper, that

> the physics and physiology of situations already give us some 'point of view' with respect to which some pairs of situations are similar in more obvious respects than others ... (Hesse ,1974, p 13)

This approach makes a difference to the way that Hesse's account of scientific inference proceeds. For example, she uses it as a 'defence' against conventionalism, whereas this book comes to a conventionalist conclusion. Hesse says of her account:

> It is not a *conventionalist* account, if by that we mean that any law can be assured of truth by sufficiently meddling with the meanings of its predicates. Such a view does not take seriously the systematic character of laws, for it contemplates preservation of the truth of a given law irrespective with its coherence with the rest of the system, that is, the preservation of simplicity and the other desirable internal characteristics of the system. Nor does it take account of the fact that not all primary recognitions of empirical similarity can be over-ridden in the interest of preserving a given law, for it is upon the existence of some such recognitions that the whole possibility of language with empirical reference rests. The present account on the other hand demands both that laws shall remain connected in an economical and convenient system and that at least most of its predicates shall remain applicable, that is, that they shall continue to depend for applicability upon the primary recognitions of similarity and difference in terms of which they were learned. (p 16)

Thus it seems that in Hesse's scheme it is physics and physiology (which give rise to primary recognitions of similarity and difference) that will prevent radical and wholesale changes in the network of law and observation. Also, it seems that this restriction will prevent the acceptance of certain laws where such acceptance requires wholesale meddling with the rest of the system. This restriction is imposed by the necessity of preserving the coherence, economy and convenience of the system along with the maintenance of a core (albeit a potentially shifting core) of primary recognitions of similarity and difference. While I agree with Hesse about the impossibility of wholesale change (see Chapter Six) the source of the continuity — more properly, locally perceived continuity — is not physics and physiology, nor logical or probabilistic coherence, nor economy and simplicity in the network of concepts, but interests and social conventions. Perceptions of probability, coherence, economy and simplicity are, in any case, themselves conventional.

Barry Barnes (1981), drawing on Hesse, makes a related point in discussing the learning situation:

> For example, Te[acher] might point to a succession of particular birds, and on each occasion say 'bird'. As a result, we might expect that L[earner] would become familiar with a number of accepted instances of 'bird', and would himself take those instances as instances of 'bird'. Such an expectation could of course be called into question on a number of grounds. It assumes, for example, that particulars can be identified in the environment, that we inherently perceive the environment as differentiated or lumpy, and that an interaction can, as it were, be brought into focus on some lump or particular. And it assumes, further, that the association of a particular with a term results in the particular being taken as an instance of the term. These are far reaching assumptions, but they are neither implausible nor readily avoidable, and will be accepted without further ado here. Essentially, what they imply is the existence of a perceptual and cognitive apparatus with at least some rudimentary inherent properties which make learning possible (p 3).

Barnes' assumptions, in my view, are only necessary where primitive learning situations are imagined. But even if they are necessary in some basic sense, it is important to forget about them. If they are important, then the only sensible way to proceed is to discover what the limits on perception are, in the world we live in *now*. However, it is not entirely clear that Barnes thinks that the 'lumpiness' of the world does place limits on what we can see. He says:

> It might be that some array of [sets of instances of terms] and generalisations could be constructed, which was inoperable as a basis for communication . . . But whether or not such a network could be constructed is of no relevance here, even though it might constitute an interesting question for philosophers. Hesse nets have here been defined as models of the verbal components of the culture of an existing community. Any existing network is known to be workable because coherent activity of some community is modelled on it (p 31).

This view appears similar to a view expressed earlier by Barnes (in his 1974) and discussed critically in Collins & Cox (1976, 1977). The view seems similar to that of Bloor (1976) who wrote:

> In particular the sociologist will be concerned with beliefs which are taken for granted or institutionalised, or invested with authority by groups of men (p 3).

Thus Barnes and Bloor appear to avoid the question of the limits potential networks imposed by the 'lumpiness' of the world, by the trick of considering only such networks as are already institutionalized. This is too 'retrospective' a view; it would be appropriate only in a static society. The trick will not work here because attempts to establish *new* inductive generalizations are examined. Thus, if it is the case that there is a class of attempts that would be bound to fail because they would render the network (or part of it) 'inoperable as a basis for communication', it would be crucial to know if any particular attempt under examination belonged to that class. For example, it might be posited that plant perception (see Chapter Five) was not institutionalized, not because no one succeeded in fitting it into the existing network of scientific institutions (essentially the view put forward here), but because there was something in the physics of the experiment and in the physiology of potential upholders of positive views about plant perception, which made it not even *potentially* institutionalizable. This would render my research otiose. Some clear guidelines are needed.

In this work the answer to the question 'how do we come to make new generalizations?' is assumed at the outset to be of the same type whether the subject about which such generalizations were being made were, say, the colour of emeralds or, say, the number of angels on the head of a pin. This is the methodological assumption. I therefore avoid talk about description; talk about description demands assumptions that talk about order avoids.

The assumptions made here are minimal. Hesse's reservations based on simplicity and coherence of the network are avoided because these concepts themselves are taken to be conventional. The links in networks are to do with taken for granted realities; coherence and simplicity cannot exist outside of a form of life. Likewise, physics and physiology are taken to play no part in the maintenance of conceptual order.

Discussion of 'primitive lumpiness', which might be necessary in order to conceive of how any generalization got under way in some mythical 'first instance', is avoided by accepting that we simply have no conceptual apparatus for thinking otherwise about mankind than as embedded in institutional/conceptual networks. It is not that we encounter mankind already institutionalized, it is that we cannot conceive of any other possibility. To quote Black (1970) '. . . the question 'Why should we accept *any* inductive rules?' can be shown to make no sense'(p 89). Thus the fact of his institutionalization is a constitutive not empirical feature of any discussion of mankind's inductive abilities. Discussion of imaginary pre-institutional situations is avoided and we are not invited to wonder if any part of our current institutionalization expresses limits on institutionalizability.

This allows us to take the principle of symmetry (Bloor, 1973) to its conclusion; all description-type language should be treated at the outset as though it did not describe anything real. This does not pose a problem for intersubjectivity because mutual understanding seems to be possible even where nothing real is the subject matter. The quality of a poem or a picture, the number of angels that could dance on the head of a pin, or the cut of the emperor's new clothes can all be discussed without there being any lumps in the world that correspond to them. Many scientists would suggest that the whole science of, for example, parapsychology is based on nothing more tangible than the emperor's clothes. But in order to take the symmetry principle seriously, discussion about the emperor's clothes must be treated as a paradigm of seeing rather than as the paradigmatic counter example. This, it must be stressed, is not a conclusion that arises out of the argument of this book. Nor is it an *a priori* epistemological claim. That kind of epistemology is not the purpose of the work. It is, what one might call a 'meta-methodological presupposition'. It is the appropriate frame of mind for doing the sociology of knowledge because it leads to the right kind of methododology. (For further discussion see Collins, 1981d.)

References Cited

Adams, D. (1979) *The Hitch Hikers Guide to the Galaxy*, London: Pan Books.

Atkinson, R. and Delamont, S. (1977) 'Mock-ups and Cock-ups: The Stage Management of Guided Discovery Instruction', in Woods, P. and Hammersley, M. (eds) *School Experience: Explorations in the Sociology of Education*, London: Croom Helm.

Backster, C. (1968) 'Evidence of a Primary Perception in Plant Life', *International Journal of Parapsychology*, X: 329–48.

Balanovski, E. and Taylor, J. G. (1978) 'Can Electromagnetism Account for Extrasensory Perception?', *Nature*, 276: 64–7.

Barber, B. (1961) 'Resistance by Scientists to Scientific Discovery', *Science*, 134: 596–602.

Barnes, S.B. (1974) *Scientific Knowledge and Sociological Theory*, London: Routledge and Kegan Paul.

Barnes, S.B. (1976) 'Natural Rationality: A Neglected Concept in the Social Sciences', *Philosophy of the Social Sciences*, 6: 115–26.

Barnes, S.B. (1981) 'On the "Hows" and "Whys" of Cultural Change (Response to Woolgar)', *Social Studies of Science*, 11: 481–98.

Barnes, S.B. (1983a) 'On the Conventional Character of Knowledge and Cognition' in Knorr-Cetina and Mulkay (1983).

Barnes, S.B. (1983b) 'Social Life as Bootstrapped Induction', *Sociology*, 4: 524-45.

Barnes, B. and Edge, D. (1982) *Science in Context: Readings in the Sociology of Science*, Milton Keynes: Open University Press.

Beloff, J. (1982) 'Die Fingerabdrucke von Psi', *Zeitschrift fur Parapsychologie*, 24: 13–24.

Beloff, J. and Bate, D. (1971) 'An Attempt to Replicate the Schmidt Findings', *Journal of the Society for Psychical Research*, 3: 21–31.

Berger, P.L. (1963) *Invitation to Sociology*, Garden City, New York: Anchor Books; London: Penguin.

Berger, P.L. and Luckman, T. (1967) *The Social Construction of Reality*, London: Allen Lane.

Berger, T. (1978) *Little Big Man*, New York: Fawcett.

Bhaskar, R. (1975) *A Realist Theory of Science*, Leeds: Leeds Books.

Black, M. (1970) *Margins of Precision: Essays in Logic and Language*, Ithaca and London: Cornell University Press.

Bloor, D. (1976) *Knowledge and Social Imagery*, London: Routledge and Kegan Paul.

Bloor, D. (1978) 'Polyhedra and the Abominations of Leviticus', *British Journal for the History of Science*, 11: 245–72.

Bloor, D. (1983) *Wittgenstein: A Social Theory of Knowledge*, London: Macmillan.

Boden, M. (1977) *Artificial Intelligence and Natural Man*, Brighton: Harvester.

Borges, J. (1970) *Labyrinths*, Harmondsworth: Penguin.

Bourdieu, P. (1975) 'The Specificity of the Scientific Field and the Social Conditions of the Progress of Reason', *Social Science Information* 14: 19–47.

Bowden, G. (1985) 'The Social Construction of Validity in Estimates of US Crude Oil Reserves', *Social Studies of Science*, 15: 2.

Brannigan, A. (1981) *The Social Basis of Scientific Discoveries*, New York: Cambridge University Press.

Braude, S. (1980) *ESP and Psychokinesis: A Philosophical Examination*, Philadelphia: Temple University Press.

Chedd, G. (1975) 'AAAS Takes on Emotional Plants', *New Scientist*, 13: 400–401.

Chubin, D.E. (1982) 'Collins's Programme and the "Hardest Possible Case"', *Social Studies of Science* 12:136–9.

Collingwood, R.G. (1946) *The Idea of History*, Oxford: Oxford Univeristy Press.

Collins, H.M. (1974) 'The TEA Set: Tacit Knowledge and Scientific Networks', *Science Studies*, 4: 165–86 (reprinted in Barnes and Edge, 1982).

Collins, H.M. (1975) 'The Seven Sexes: A Study in the Sociology of a Phenomenon, or the Replication of Experiments in Physics', *Sociology*, 9: 205–24 (reprinted in Barnes and Edge, 1982).

Collins, H.M. (1976) 'Upon the Replication of Scientific Findings: A Discussion Illuminated by the Experiences of Researchers into Parapsychology', *Proceedings of 4S/ISA Conference*, Cornell University.

Collins, H.M. (1979) 'The Investigation of Frames of Meaning in Science: Complementarity and Compromise', *Sociological Review*, 27: 703–18.

Collins, H.M. (ed) (1981a) *Knowledge and Controversy: Studies of Modern Natural Science* special issue of *Social Studies of Science*, 11: 1.

Collins, H.M. (1981b) 'Stages in the Empirical Programme of Relativism', in Collins, (1981a): 3–10.

Collins, H.M. (1981c) 'Son of the Seven Sexes: The Social Destruction of a Physical Phenomenon', in Collins (1981a): 33–62

Collins, H.M. (1981d) 'What is TRASP? The Radical Programme As A Methodological Imperative', *Philosophy of the Social Sciences*, 11: 215–24.

Collins, H.M. (1982a) 'Special Relativism–The Natural Attitude', *Social Studies of Science* 12: 139–43.

Collins, H.M. (1982b) *Sociology of Scientific Knowledge: A Sourcebook*, Bath: Bath University Press.

Collins, H.M. (1983a) 'The Meaning of Lies: Accounts of Action and Participatory Research', in Gilbert, G. N. and Abell, P. (eds), *Accounts and Action: Surrey Conferences on Sociological Theory and Method 1*, Aldershot: Gower.

Collins, H.M. (1983b) 'Scientific Knowledge and Science Policy: Some Foreseeable Implications', presented at the 1983 Annual Meeting of the Society for Social Studies of Science, Blacksburg, Virginia. Part published in *European Association for the Study of Science and Technology Newsletter*, 2: 5–8.

Collins, H.M. (1983c) 'The Sociology of Scientific Knowledge: Studies of Contemporary Science', *Annual Review of Sociology*, 9: 265–85.

Collins, H.M. (1984a) 'Concepts and Practice of Participatory Fieldwork', in Bell, C. and Roberts, H. (eds), *Social Researching*, London: Routledge and Kegan Paul.

Collins, H.M. (1984b) 'When do Scientists Prefer to Vary Their Experiments?', *Studies in the History and Philosophy of Science*, 15: 169–74.

Collins, H. M. and Cox, G. (1976) 'Recovering Relativity: Did Prophecy Fail?', *Social Studies of Science*, 6: 423–44.

Collins, H.M. and Cox, G. (1977) 'Relativity Revisited: Mrs. Keech, A Suitable Case for Special Treatment?', *Social Studies of Science*, 7: 327–80.

Collins, H.M. and Harrison, R. (1975) 'Building a TEA Laser: The Caprices of Communication', *Social Studies of Science*, 5: 441–5.

Collins, H.M. and Pinch, T. J. (1979) 'The Construction of the Paranormal: Nothing Unscientific is Happening' in Wallis, R. (ed) (1979) (reprinted in Collins, 1982b).

Collins, H.M. and Pinch, T. J. (1981) 'Rationality and Paradigm Allegiance in Extraordinary Science', in Hans Peter Duerr (ed) *The Scientist and the Irrational*, Frankfurt: Syndikat (in German).

Collins, H.M. and Pinch, T. J. (1982) *Frames of Meaning: The Social Construction of Extraordinary Science*, London: Routledge and Kegan Paul.

Collins, H.M. and Shapin, S. (1984) 'The Historical Role of the Experiment', Proceedings of the International Conference on Using History of Physics in Innovatory Physics Education, Pavia, Italy, 5–9 September 1983. (A shortened version was published as 'Uncovering the Nature of Science', *Times Higher Education Supplement*, 27 July 1984: 13.)

Cotgrove, S. F. (1982) *Catastrophe or Cornucopia: The Environment, Politics and the Future*, London and New York: John Wiley and Sons.

Critchley, O. H. (1978) 'Aspects of the Historical, Philosophical and Mathematical Background to the Statutory Management of Nuclear Plant Risks in the United Kingdom' pp 11–18 in *Radiation Protection in Nuclear Power Plants and the Fuel Cycle*, London: BNES.

Davies, P.C.W. (1980) *The Search for Gravity Waves*, Cambridge: Cambridge University Press.

Dennis, M. (1985) 'Drilling for Dollars: The Making of American Petroleum Reserve Estimates, 1921-25', *Social Studies of Science*, 15: 2.

Dreyfus, H. (1979) *What Computers Can't Do*, New York: Harper and Row.

Farley, J. and Geison, G. L. (1974) 'Science Politics and Spontaneous Generation in Nineteenth-Century France: The Pasteur-Pouchet Debate', *Bulletin of the History of Medicine*, 48: 161–98 (reprinted in Collins 1982b).

Feyerabend, P. K. (1975) *Against Method*, London: New Left Books.

Fleck, Ludwik (1979) *Genesis and Development of a Scientific Fact*, Chicago: University of Chicago Press (first published in German in 1935).

Frankel, E. (1976) 'Corpuscular Optics and the Wave Theory of Light: The Science and Politics of a Revolution in Physics', *Social Studies of Science*, 6: 141–84.

Franklin, A. and Howson, C. (1984) 'Why do Scientists Prefer to Vary Their Experiments?', *Studies in the History and Philosophy of Science*, 15: 51–62.

Friedman, N. (1967) *The Social Nature of Psychological Research*, New York: Basic Books.

Garfinkel, H. (1967) *Studies in Ethnomethodology*, New Jersey: Prentice-Hall.

Gellner, E. (1974) 'The New Idealism: Cause and Meaning in the Social Sciences' in Giddens (ed) *Positivism and Sociology*, London: Heinemann.

Gier, N. F. (1981) *Wittgenstein and Phenomenology*, Albany: State University of New York Press.

Gieryn, T. (1983) 'Boundary-Work and the Demarcation of Science from Non-Science: Strains and Interests in Professional Ideologies of Scientists', *American Sociological Review*, 48: 781–95.

Gillespie, B., Eva, D. and Johnston, R. (1979) 'Carcinogenic Risk Assessment in the United States and Great Britain: The Case of Aldrin/Dieldrin' *Social Studies of Science*, 9: 265–301.

Gooding, D.G. (1986) 'How Do Scientists Reach Agreement About Novel Observations?', *Studies in the History and Philosophy of Science*, 17, 1986 (forthcoming).

Goodman, N. (1973) *Fact, Fiction, and Forecast* (3rd Edition), New York: Bobbs-Merrill.

Goodman, N. (1978) *Ways of Worldmaking*, Indianapolis: Hacket.

Hacking, I. (1983) *Representing and Intervening: Introductory Topics in the Philosophy of the Natural Sciences*, Cambridge: Cambridge University Press.

Harvey, B. (1981) 'Plausibility and the Evaluation of Knowledge: A Case Study in Experimental Quantum Mechanics', in Collins, 1981a: 95–130.

Henkel, R.E. and Morrison, D.E. (1970) *The Significance Test Controversy*, London: Butterworths.

Hesse, M. (1974) *The Structure of Scientific Inference*, London: Macmillan.

Holton, G. (1978) *The Scientific Imagination*, Cambridge: Cambridge University Press.

Horowitz, *et al* (1975) 'Plant Primary Perception: Electrophysiological Unresponsiveness to Brine Shrimp Killing', *Science* 189: 478–80

Johnson, R. (1971) 'The Influence of Temperature and Humidity on the Low Frequency Capacitance and Conductance Across a Philodendron Leaf (A Study of the Backster Effect)', Thesis in Partial Fulfilment of M.Sc. in Electrical Engineering, University of Washington.

Johnson, R. (1972) 'To the Editors', *The Journal of Parapsychology*, 36: 71–2.

Knorr-Cetina, K.D. (1981) *The Manufacture of Knowledge*, Oxford: Pergamon Press.

Knorr-Cetina, K.D. and Cicourel, A. (eds) (1981) *Advances in Social Theory and Methodology: Toward an Integration of Micro- and Macro-Sociologies*, London: Routledge and Kegan Paul.

Knorr-Cetina, K.D. and Mulkay, M. (eds) (1983) *Science Observed: Perspectives on the Social Study of Science*, London: Sage.

Kuhn, T.S. (1962) *The Structure of Scientific Revolutions*, Chicago: University of Chicago Press.

Lakatos, I. (1970) 'Falsification and the Methodology of Scientific Research Programmes', in Lakatos and Musgrave (1970).

Lakatos, I. (1976) *Proofs and Refutations*, Cambridge: Cambridge University Press.

Lakatos, I. and Musgrave, A. (eds) (1970) *Criticism and the Growth of Knowledge*, Cambridge: Cambridge University Press.

Langmuir, I. (1953) (revised by R. N. Hall, 1968), 'Pathological Science', *General Electric R and D Centre Report*, Number 68–C–035, New York.

Latour, B. (1983) 'Give me a Laboratory and I Will Raise the World' in Knorr-Cetina and Mulkay (1983).

Latour, B. and Woolgar, S. (1979) *Laboratory Life: The Social Construction of Scientific Facts*, London and Beverly Hills: Sage.

Laudan, L. (1982) 'A Note On Collins's Blend of Relativism and Empiricism', *Social Studies of Science*, 12: 131–32.

Lynch, M., Livingstone, E. and Garfinkel, H. (1983) 'Temporal Order in Laboratory Work', in Knorr-Cetina and Mulkay (1983).

Mackenzie, D. (1981) *Statistics in Britain 1865-1930*, Edinburgh: Edinburgh University Press.

Markle, G.E. and Petersen, J.C. (eds) (1980) *Politics, Science and Cancer: The Laetrile Phenomenon*, Boulder, Colorado: Westview Press.

Mazur, A. (1981) *Dynamics of Technical Controversy*, Washington DC: Communications Press.

McCorduck, P. (1979) *Machines Who Think: A Personal Inquiry Into the History and Prospects of Artificial Intelligence*, San Francisco: Freeman.

Merton, R.K. (1973) 'Paradigm for the Sociology of Knowledge', as reprinted in *The Sociology of Science: Theoretical and Empirical Investigations*, Chicago: Chicago University Press.

Mulkay, M. Potter, J. and Yearley, S. (1983) 'Why an Analysis of Scientific Discourse Is Needed', in Knorr-Cetina and Mulkay (1983).

Munson, T.N. (1963) 'Wittgenstein's Phenomenology', *Philosophy and Phenom-enological Research*, 37–50.

Myers, G. (1985a) 'The Social Construction of Two Biology Articles' *Social Studies of Science* 15: (forthcoming).

Myers, G. (1985b) 'The Social Construction of Two Biologists' Proposals' *Written Communication*, 1, 3: (forthcoming).

Nelkin, D. (1975) 'The Political Impact of Technical Expertise', *Social Studies of Science*, 5: 35–54.

Nelkin, D. (1978) 'Threats and Promises: Negotiating the Control of Research', *Daedelus*, 107: 191–209.

Nelkin, D. (ed.) (1979) *Controversy: Politics of Technical Decisions*, Beverly Hills: Sage.

Nickles, T. (1980) *Scientific Discovery: Case Studies*, Boston Studies in the Philosophy of Science, No. 60, Dordrecht: Reidel.

O'Brien, F. (1974) *The Third Policeman*, London: Picador.

Oteri, Weinberg and Pinales (1978) 'Cross Examination of Chemists in Narcotics and Marijuana Cases', *Contemporary Drug Problems*, 2: 225–38 (reprinted in Barnes and Edge, 1982).

Overington, M. A. (1979) 'Doing the What Comes Rationally: Some Developments in Metatheory', *The American Sociologist*, 14: 2–12.

Peter J.P. and Olson J.C. (1983) 'Is Science Marketing?', *Journal of Marketing* 57, 111–25.

Petersen, J.C. and Markle, G.E. 'Politics and Science in the Laetrile Controversy', *Social Studies of Science*, 9: 139–66.

Pickering, A. R. (1980) 'The Role of Interests in High-Energy Physics: The Choice Between Charm and Colour' pp 107–38 in Knorr, K. D., Krohn, R. and Whitley, R. (eds) *The Social Process of Scientific Investigation. Sociology of the Sciences*, IV, Dordrecht: Reidel.

Pickering, A. (1981a) 'Constraints on Controversy: The Case of the Magnetic Monopole', in Collins, 1981a: 63–93.

Pickering, A. (1981b) 'The Hunting of the Quark', *ISIS*, 72: 216–36.

Pickering, A. (1984) *Constructing Quarks: A Sociological History of Particle Physics*, Edinburgh: Edinburgh University Press.

Pinch, T. J. (1977) 'What Does a Proof Do If It Does Not Prove?', in Mendelsohn, E., Weingart, P. and Whitley, R., *The Social Production of Scientific Knowledge*, Dordrecht: Reidel.

Pinch, T. J. (1981) 'The Sun-Set: The Presentation of Certainty in Scientific Life', in Collins, 1981a: 131–58.

Pinch, T. J. (1985) 'Towards an Analysis of Scientific Observation: The Externality and Evidential Significance of Observation Reports in Physics', *Social Studies of Science* 15: 1.

Pinch, T.J. (forthcoming) *Confronting Nature*, Dordrecht: Reidel.

Pinch T.J. and Bijker, W. (1984) 'The Social Construction of Facts and Artefacts: A Unified Approach Toward the Study of Science and Technology', *Social Studies of Science*, 14: 399–42.

Pinch, T.J. and Collins, H.M. (1984) 'Private Science and Public Knowledge: The Committee for the Scientific Investigation of the Claims of the Paranormal and Its Use of the Literature', *Social Studies of Science*, 14: 521–46.

Polanyi, M. (1958) *Personal Knowledge*, London: Routledge and Kegan Paul.

Polanyi, M. (1962) 'The Republic of Science, Its Political and Economic Theory', *Minerva*, 1: 54–73.

Polanyi, M. (1967) *The Tacit Dimension*, New York: Anchor.

Popper, K. R. (1959) *The Logic of Scientific Discovery*, New York: Harper & Row.

Ravetz, J.R. (1971) *Scientific Knowledge and Its Social Problems*, Oxford: Oxford University Press.

Rhine, J. B. (1971) 'News and Comments', *The Journal of Parapsychology*, 35: 247.

Robbins, D. and Johnston, R. (1976) 'The Role of Cognitive and Occupational Differentiation in Scientific Controversies', *Social Studies of Science*, 6: 349−68 (reprinted in Collins 1982b).

Roche, M. (1973) *Phenomenology, Language and the Social Sciences*, London: Routledge and Kegan Paul.

Roll-Hansen, N. (1979) 'Experimental Method and Spontaneous Generation: The Controversy Between Pasteur and Pouchet, 1859−64', *Journal of the History of Medicine and Allied Sciences*, 34: 273−92.

Rosenthal, R. (1978) 'Interpersonal Expectancy Effects: The First 345 Studies', *The Behavioural and Brain Sciences*, 3: 377−415.

Rosenthal, R. (1969) 'Interpersonal Expectations', in Rosenthal, R. and Rosnow, R. C. (eds) *Artifacts in Behavioural Research*, New York: Academic Press.

Schmeidler, G. R. and McConnell, R. A. (1958) *ESP and Personality Patterns*, New Haven: Yale University Press.

Schmidt, H. (1969a) 'Quantum Processes Predicted?', *New Scientist*, 16 October: 114−5.

Schmidt, H. (1969b) 'Precognition of a Quantum Process', *The Journal of Parapsychology*, 33: 99−108.

Schmidt, H. (1970) 'Quantum Mechanical Random-Number Generator', *Journal of Applied Physics*, 41: 2.

Schutz, A. (1962), *The Problem of Social Reality, Collected Papers*, Vol. I, The Hague: Martinus Nijhoff.

Schutz, A. (1964) *Studies in Social Theory, Collected Papers*, Vol. II, The Hague: Martinus Nijhoff.

Shapin, S. (1979) 'The Politics of Observation: Cerebral Anatomy and Social Interests in the Edinburgh Phrenology Disputes', in Wallis (1979) (reprinted in Collins, 1982b).

Shapin, S. (1984) 'Pump and Circumstances: Robert Boyle's Literary Technology', *Social Studies of Science*, 14: 481−520.

Shapin, S. and Schaffer, S. (1985) *Leviathan and the Air Pump: Hobbes, Boyle and the Experimental Life*, Princeton: Princeton University Press.

Specht, E. K. (1969) *The Foundations of Wittgenstein's Late Philosophy*, Manchester: Manchester University Press (English edition).

Spiegelberg, H. (1959) 'How Subjective is Phenomenology?', *Proceedings for the Year of 1959 of the American Catholic Philosophical Association*, 28−36.

Spiegelberg, H. (1969) *The Phenomenological Movement: A Historical Introduction* (two volumes), The Hague: Martinus Nijhoff.

Studer, K.E. and Chubin, D.E. (1980) *The Cancer Mission; Social Contexts of Biomedical Research*, Beverly Hills: Sage.

Tart, C. (1973) 'Parapsychology', *Science*, 182: 222.

Taylor, C. and Ayer, A. J. (1959) 'Phenomenology and Linguistic Analysis' *Aristotelian Society, Supplementary Volume*, 33: 93−124.

Taylor, J.G. (1971) *The Shape of Minds To Come*, New York: Weybright and Talley.

Taylor, J.G. (1975) *Superminds: An Enquiry Into the Paranormal*, London Macmillan.

Taylor, J.G. (1980) *Science and the Supernatural*, London: Temple Smith.

Taylor, J.G. and Balanovski, E. (1979) 'Is There Any Scientific Explanation of the Paranormal?', *Nature*, 279: 631–3.

Tompkins, P. and Bird, B. (1974) *The Secret Life of Plants*, London: Allen Lane.

Travis, G. D. L. (1981) 'Replicating Replication? Aspects of the Social Construction of Learning in Planarian Worms', in Collins, (1981a): 11–32.

Van Peursen, C. A. (1959) 'Edmund Husserl and Ludwig Wittgenstein', *Philosophy and Phenomenological Research*, XX: 181–97.

Vonnegut, Kurt (1979) *Slaughterhouse 5*, London: Panther.

Wallis, R. (ed.) *On the Margins of Science: The Social Construction of Rejected Knowledge, Sociological Review Monograph* 27, Keele: University of Keele Press.

Wilson, Bryan R. (ed.) (1970) *Rationality*, Oxford: Blackwell.

Wilson T. P. (1970) 'Normative and Interpretative Paradigms in Sociology' in Douglas, J. (ed) (1970) *Understanding Everyday Life*, London: Routledge and Kegan Paul.

Winch, P. (1958) *The Idea of a Social Science*, London: Routledge and Kegan Paul.

Wittgenstein, L. (1953) *Philosophical Investigations*, Oxford: Blackwell.

Wolstenholme, and Miller (eds) (1956) *CIBA Foundation Symposium on Extrasensory Perception*, London: J. and A. Churchill Ltd.

Wynne, B. (1982) *Rationality or Ritual? The Windscale Inquiry and Nuclear Decisions In Britain*, Chalfont St. Giles, Bucks: British Society for the History of Science Monograph.

Yearley, S. (1984) *Science and Sociological Practice*, Milton Keynes: Open University Press.

Zuckerman, H. A. (1977) 'Deviant Behavioural Social Control in Science', in Sagarin, E. (ed) *Deviance and Social Change*, Beverly Hills: Sage.

Name Index

Subject Index